OXFORD REFERENCE

A Dictionary of Dates

Cyril Leslie Beeching was a professional musician and entertainer for fifteen years, writing songs, scripts and other material for broadcasting and recording. In more recent years he has written books, notably *A Dictionary of Eponyms* (also in Oxford Reference), and has contributed to newspapers such as the *Observer* and *Guardian*, mostly on his specialist subject of dates and anniversaries, about which he has given broadcasts and supplied information to television and radio programmes.

A
DICTIONARY
OF DATES

Cyril Leslie Beeching

Oxford   New York
OXFORD UNIVERSITY PRESS
1993

Oxford University Press, Walton Street, Oxford OX2 6DP

Oxford New York Toronto
Delhi Bombay Calcutta Madras Karachi
Kuala Lumpur Singapore Hong Kong Tokyo
Nairobi Dar es Salaam Cape Town
Melbourne Auckland Madrid

and associated companies in
Berlin Ibadan

Oxford is a trade mark of Oxford University Press

First published 1993 as an Oxford University Press paperback

British Library Cataloguing in Publication Data
Data available

Library of Congress Cataloging in Publication Data
Beeching, Cyril Leslie.
A dictionary of dates / Cyril Leslie Beeching.
p. cm. Includes index.
1. Anniversaries—Dictionaries. 2. Chronology, Historical.
3. Calendars. I. Title.
902'.02—dc20 D11.5.B37 1993 92–40815
ISBN 0-19-285274-4

10 9 8 7 6 5 4 3 2 1

Typeset by Oxuniprint and
Latimer Trend & Co. Ltd.
Printed in Great Britain by
Biddles Ltd.
Guildford and King's Lynn

To Frank Atkinson,
latterly Librarian of
St Paul's Cathedral,
in remembrance of his
help and encouragement
in the compilation of
this book

CONTENTS

INTRODUCTION

In the preparation of this book, it has become clear that most readers have an interest in dates in one way or another: from the obvious personal anniversaries to the wider awareness of historical events of lesser or greater significance. It is also undoubtedly the information about the people and happenings *behind* the dates which gives them their particular attraction and makes their listing so much more than a straightforward catalogue or directory. (There are, too, a remarkable number of fascinating coincidences concerning birthdays and other occurrences on certain days of the year, though no claims are made for any astrological significance in this respect.)

The book's two main sections are intended to give easy references to the dates and their associated details in both day-by-day and chronological order. In addition there is a very extensive index which, apart from its main function, could be seen as a 'good read' in itself.

Inevitably, if only for reasons of finite space, there are some names and items which the reader will look for in vain; but it has been the intention to cover as wide a field of interest as possible. Some items, however, have been deliberately omitted because of the lack of certainty about the dates; and in the instances of dates being duplicated through the introduction of the Gregorian calendar, the usual convention of taking the New Style has generally been followed. (For fuller details of this aspect the reader may refer to 'Gregorian calendar' in the index of this book.)

C.L.B.

New Malden, Surrey
January 1993

JANUARY ❧

THIS day was first recognized as New Year's Day with the introduction of the Gregorian calendar in France, Italy, Portugal, and Spain in 1582; yet it was not until 1752 that the New Style was adopted in Britain.

The oldest existing London daily newspaper, *The Times*, was first published on this day in 1788, replacing the *Daily Universal Register*, which had been in circulation for three years; and in 1814 the first Welsh-language newspaper, *Seren Gomer*, was published in Swansea.

Born on this day were: Baron Pierre de Coubertin (1863–1937), the French sportsman who revived the Olympic Games in Greece; the English novelist E(dward) M(organ) Forster (1879–1970); J(ohn) Edgar Hoover (1895–1972), the director of the FBI from 1924 until his death; Harold Adrian Russell 'Kim' Philby (1912–88), the British double agent known as 'the third man'; and the British playwright Joe Orton (1933–67).

In 1909 the first payments of old age pensions in Great Britain were made, with all persons over the age of 70 with an annual income of less than £21 receiving five shillings a week.

❧ 2 January

ON this day . . . in 1839 the French photographic pioneer Louis Daguerre took the first photograph of the moon (in Paris); and in 1959 the Russian artificial satellite *Luna I*, the first rocket to pass the moon, was launched from Tyuratam.

Born on this day were: General James Wolfe (1727–59), the British soldier and hero of the battle of Quebec; Gilbert Murray (1866–1957), the Australian-born scholar who became the chairman of the League of Nations 1923–38; Arthur William Charles 'Wentworth' Gore (1868–1928), the English tennis player who appeared at Wimbledon on every occasion from 1888 to 1927 and was the men's singles champion in 1901, 1908, and

1909, when he became the oldest winner of the championship; and the English composer Sir Michael Tippett (1905–).

℅ 3 January

ON this day ... in 1911 what came to be known as 'the Sidney Street Siege' was witnessed by two future British prime ministers: one of them was Clement Attlee, whose 28th birthday it happened to be; and the other was Winston Churchill, at that time the Home Secretary and personally involved in the proceedings. (The incident was a sequel to 'the Houndsditch murders' of 16 December, in which three police officers had been shot, and the siege ended when the house was burned down with its occupants as the police and the military were attempting to storm the building.)

Born on this day were: Herbert Morrison, later Lord Morrison of Lambeth (1888–1965), who was a cabinet minister in both the Churchill and Attlee administrations; John Ronald Reuel Tolkien (1892–1973), the philologist and author of the mythological tales *The Hobbit* and *The Lord of the Rings*; and the Danish-born pianist and entertainer Victor Borge (1909–).

William Joyce, the British traitor who had broadcast for the Nazi regime in World War II, known as 'Lord Haw-Haw' from his affected 'upper-class' drawl, was hanged at Wandsworth Prison in 1946.

In 1959 Alaska was admitted to the Union as the forty-ninth state.

℅ 4 January

BORN on this day were: Jacob Grimm (1785–1863), the German philologist who invented the word 'umlaut' (but is best known for his *Fairy Tales*, written with his brother Wilhelm); Louis Braille (1809–52), the Frenchman who invented and gave his name to the universal system of reading and writing for the blind (and was himself blind from the age of 3); and Sir Isaac Pitman (1813–97), the Englishman who invented the shorthand system named after him.

Also born on this day were: the Welsh painter Augustus John (1878–1961); Jane Wyman (1914–), the American actress who was the first wife of the US president Ronald Reagan; the American Soprano Grace Bumbry (1937–); and the American boxer Floyd Patterson (1935–), who was the youngest world heavyweight champion when he won the title in 1956 and the first to regain the title, in 1960.

In 1938 the circus was first televised with Bertram Mills' Circus transmitted live from Olympia. (And since this was also the first time that pictures were shown of the paying public at any event, the customers were informed that they could, if they so desired, be seated out of the cameras' range!)

✺ 5 January

BORN on this day were: the Italian pianist Arturo Benedetti Michelangeli (1920–); the Austrian-born pianist Alfred Brendel (1931–), who is an honorary KBE; and the Italian pianist Maurizio Pollini (1942–).

Also born on this day were: King Camp Gillette (1855–1932), the American inventor of the safety razor; Konrad Adenauer (1876–1967), the first chancellor of West Germany, in 1949; HRH Jean, the Grand Duke of Luxembourg (1921–); Zulfikar Ali Bhutto (1928–79), the president of Pakistan who was executed for conspiring to murder a political rival; HM King Juan Carlos of Spain (1938–); and Mansur Ali Khan, formerly the Nawab of Pataudi (1941–), the former Indian cricket captain.

The Antarctic explorer Sir Ernest Shackleton died at South Georgia on his fourth and last expedition in 1920–2; and in 1941 the pioneer airwoman Amy Johnson was drowned after baling out of her aircraft over the Thames Estuary.

✺ 6 January

IN 1540 Henry VIII was married to his fourth wife Anne of Cleves. (The marriage was declared null and void after some six months: 'The

King found her so different from her picture ... that ... he swore they had brought him a Flanders mare.')

Born on this day were: Richard II (1367–1400), the eighth and last of the Plantagenet kings of England; Percival Pott (1714–88), the English surgeon who gave his name to a fracture and a disease; Gustave Doré (1832–83), the French painter and illustrator; the German composer Max Bruch (1838–1920); the Russian composer Alexander Scriabin (1872–1915), whose nephew was the Russian revolutionary leader Molotov; and Tom Mix (1880–1940), the Texas ranger and US marshal who became one of the best-known film actors in 'Westerns' with his horse 'Tony'.

೫ 7 January

ON this day ... in 1558 Calais was finally taken from the British by the French, under the Duke of Guise; and the first Channel crossing by air was made from Dover to Calais by the Frenchman Jean-Pierre Blanchard and the American Dr John Jeffries in a hydrogen balloon, in 1785.

Born on this day were: Jacques Étienne Montgolfier (1745–99), the French inventor who with his brother Joseph Michel Montgolfier built the first successful hot-air balloon, in 1783; the German inventor Johann Philipp Reis (1834–74), who first demonstrated an electrical telephone in 1861 (fifteen years before Bell's invention was patented); and Arthur Clifford Hartley (1889–1960), the English inventor of World War II's PLUTO (Pipeline under the Ocean) and FIDO (Fog Investigation Dispersal Operation).

Also born on this day were: Millard Fillmore (1800–74), the thirteenth US president, who came to office on the death of Zachary Taylor, in 1850; Adolph Zukor (1873–1976), the Hungarian-born pioneer of the film industry and its first centenarian; and the French composer Francis Poulenc (1899–1963).

❦ 8 January

Elvis Presley (1935–77), the American singer known as 'the king of rock 'n' roll', was born on this day: and two other popular singers, Shirley Bassey (1937–   ) and David Bowie (1947–   ), share his birthday.

Also born on this day were: Wilkie Collins (1824–89), the English novelist who introduced the first detective in fiction, 'Sergeant Cuff', in his novel *The Moonstone*; and James Craig, the first Viscount Craigavon (1871–1940), the first prime minister of Northern Ireland.

❦ 9 January

BORN on this day were: John Jervis, the Earl of St Vincent (1735–1823), the British admiral who took his title from the third battle of St Vincent, in 1797; Jeanette 'Jennie' Jerome, Lady Randolph Churchill (1854–1921), Sir Winston Churchill's American-born mother; Karel Čapek (1890–1938), the Czech writer who introduced the word 'robot' into the language, in his play *RUR* (*Rossum's Universal Robots*); Dame Gracie Fields (1898–1979), the English singer and comedienne who became internationally famous; and the thirty-seventh US president Richard Milhous Nixon (1913–   ), who was the first to resign his office, in 1974.

In 1972 the *Queen Elizabeth*, for more than thirty years the world's largest passenger liner, was gutted by fire in Hong Kong harbour.

❦ 10 January

ON this day . . . in 1645 the archbishop of Canterbury William Laud was beheaded on Tower Hill, having been found 'guilty of endeavouring to subvert the laws, to overthrow the Protestant religion, and to act as an enemy to Parliament'. (It was to be more than fifteen years before the next archbishop was appointed, with the Restoration of Charles II.)

In 1840 the 'penny post' (the British system of delivering mail at a standard charge, regardless of distance) came into operation. This coincided with the introduction of the first correspondence course, which was for Isaac Pitman's system of shorthand; and in 1863 the first underground railway, London's Underground Metropolitan Railway, was opened to fare-paying passengers, with trains running at fifteen-minute intervals from Farringdon Street to Paddington.

Born on this day were: the historian and politician Lord Acton (1834–1902), who told us that 'Power tends to corrupt and absolute power corrupts absolutely'; Charles Adrien Wettach (1880–1959), the Swiss clown known as 'Grock'; Dame Barbara Hepworth (1903–75), the British sculptor; and Galina Ulanova (1910–   ), the Russian ballerina.

ఈ 11 *January*

ON this day . . . in 1922 insulin was first administered to a diabetic patient (a 14-year-old Canadian boy, Leonard Thompson), who went on to lead a normal life. (It was only the previous year that the Canadian physiologists Frederick Grant Banting and Charles Herbert Best had discovered and isolated the hormone.)

Another Canadian, Sir John Alexander Macdonald (1815–91), the first prime minister of Canada (in 1867), was born on this day (in Glasgow).

Also born on this day were: William James (1842–1910), the American philosopher and brother of the novelist Henry James; the Norwegian composer Christian Sinding (1856–1941), best remembered for his piano composition *Rustle of Spring*; Fred Archer (1857–86), the British jockey whose 2,749 winning mounts included five Derby winners; and Alan Paton (1903–88), the South African writer best remembered for his novel *Cry, the Beloved Country*.

ఈ 12 *January*

ON this day . . . in 1866 the Aeronautical Society of Great Britain (now the Royal Aeronautical Society) was established, fourteen years after the first such organization, the *Société Aérostatique de France*, and thirty-seven years before the first successful powered flight by the Wright Brothers, in 1903.

Jean Joseph Étienne Lenoir, the Frenchman who invented the first practical internal combustion engine (which of course made powered flight possible), was born in 1822; though he died in 1900, some three years before his invention was used to power an aeroplane.

Also born on this day were: Edmund Burke (1729–97), the British statesman and philosopher; Johann Heinrich Pestalozzi (1746–1827), the Swiss educational reformer and philanthropist; and the American painter John Singer Sargent (1856–1925).

℘ 13 January

ON this day . . . in 1910 opera was first heard on the radio on the first 'outside broadcast', with a performance of Mascagni's *Cavalleria rusticana* and Leoncavallo's *Pagliacci* from the Metropolitan Opera House, New York. (Caruso sang the role of Canio in '*Pag.*')

Born on this day were: the Australian singer Peter Dawson (1882–1961), who made the first of 3,000 or more recordings in 1904 and was also a successful songwriter (as J. P. McCall); Lord 'Ted' Willis (1918–92), the English playwright and novelist who created the best-known British bobby, 'Dixon of Dock Green'; and Michael Bond (1926–   ), the English writer for children who created 'Paddington Bear'.

℘ 14 January

BORN on this day were: Ludwig von Köchel (1800–77), the Austrian botanist and musicologist who classified and gave the initial letter of his surname to the works of Mozart; Dr Albert Schweitzer (1875–1965), the French philosopher, musician, and medical missionary who founded the Lambaréné Hospital in French Equatorial Africa; Hugh Lofting (1886–1947), the British-born writer who created 'Dr Dolittle'; Sir Cecil Beaton (1904–80), the photographer, writer, and theatrical designer who created the scenery and costumes for *My Fair Lady*, *Gigi*, etc.;

and the English actor Warren Mitchell (1926—   ), who created the role of 'Alf Garnett', the bigoted British proletarian.

✌ 15 January

BORN on this day were: William Prout (1785–1850), the English physician and chemist who anticipated the atomic theory; and Edward Teller (1908—   ), the Hungarian-born American scientist known as 'the father of the hydrogen bomb'.

In 1962, the centigrade or Celsius scale was first used in the British Meteorological Office weather forecasts, more than 200 years after the death of the Swedish scientist who invented and gave his name to the scale.

Also born on this day were: Ivor Novello (1893–1951), the Welsh composer and actor; and Martin Luther King (1929–68), the American civil rights leader and Nobel peace prize recipient for 1964.

The Irish Free State was established in 1922.

✌ 16 January

ON this day . . . in 1780 the Duke of Clarence, the third son of George III and later William IV, took part in the second battle of Cape St Vincent as a midshipman (and was subsequently known as 'the Sailor King'); and in 1809, at the battle of Corunna, in Spain, Sir John Moore was mortally wounded in his moment of victory over the French under Marshal Soult and his burial ('darkly at dead of night') is commemorated in the poem by Charles Wolfe. ('Slowly and sadly we laid him down, | From the field of his fame fresh and gory; | We carved not a line, and we raised not a stone, | But we left him alone with his glory.')

Robert Service (1874–1958), the English-born poet and novelist known as 'the Canadian Kipling', was born on this day; and the English poet Edmund Spenser, known as 'the Poets' Poet', died in 1599. (His exact date of birth is not known.)

❦ 17 *January*

ON this day . . . in 1746, at the battle of Falkirk, 'Bonnie Prince Charlie' and his Highlanders had their last victory in the Forty-five Jacobite uprising before their eventual defeat at Culloden three months afterwards; and in 1991 the 'Gulf War' began with air raids on Iraq by allied US, British, and Saudi forces to liberate Kuwait.

Born on this day were: Benjamin Franklin (1706–90), the American statesman, philosopher, and scientist (who invented the lightning-conductor); Anne Brontë (1820–49), the English poet and novelist and the youngest of the Brontë sisters of Haworth; David Lloyd George, the first Earl Lloyd-George (1863–1945), British prime minister 1916–22; David Beatty, the first Earl Beatty (1871–1936), the commander of the British battle cruisers at the battle of Jutland in 1916; the British novelist Sir Compton Mackenzie (1883–1972); the British novelist Nevil Shute (1899–1960); and Muhammad Ali (1942–   ), formerly known as Cassius Clay, the American world heavyweight boxing champion who was the first to regain the title three times.

❦ 18 *January*

ON this day . . . in 1778 Captain James Cook discovered the Hawaiian Islands, naming them the 'Sandwich Islands', after Lord Sandwich, the First Lord of the Admiralty at that time; and in 1912 Captain Robert Falcon Scott and his party reached the South Pole to discover that the expedition led by the Norwegian explorer Amundsen had reached it a month before.

Three celebrated English writers of books for children were born on this day: A(lan) A(lexander) Milne (1882–1956), creator of 'Winnie the Pooh'; Arthur Ransome (1884–1967); and Raymond Briggs (1934–   ), who also illustrates his books.

Also born on this day were: Peter Mark Roget (1779–1869), the English physician and scholar who compiled the *Thesaurus of English Words and Phrases*; Matthew Webb, known as 'Captain Webb' (1848–83), the Englishman who was the first to swim the English Channel; the first prime minister of the Australian Commonwealth, Sir Edmund Barton

(1849–1920); and Sir Thomas Octave Murdoch 'Tommy' Sopwith (1888–1989), the British aircraft designer and sportsman whose long career was involved in the evolution of the aeroplane from its earliest days to the jet age.

✛ 19 January

BORN on this day were: James Watt (1736–1819), the Scottish inventor who gave his name to a unit of power, and the English inventor Sir Henry Bessemer (1813–98), who gave his name to a process for converting cast iron into steel.

Also born on this day were: Robert E(dward) Lee (1807–70), the commander of the Confederate armies in the American Civil War; Edgar Allan Poe (1809–49), the American poet and short story writer who pioneered the modern detective story; Paul Cézanne (1839–1906), the French painter who pioneered the Impressionist movement in painting; Javier Pérez de Cuéllar (1920–   ), the Peruvian diplomat who was the secretary-general of the United Nations 1982–92; Patricia Highsmith (1921–   ), the American crime fiction writer; and the English conductor Simon Rattle (1955–   ).

✛ 20 January

ON this day . . . in 1327 Edward II was deposed by his eldest son, who became king of England as Edward III on 25 January.

In 1841 Hong Kong was ceded to Britain by China by the Chuenpi Convention. (This was confirmed by the Treaty of Nanking in the following year.)

Born on this day were: the American comedian George Burns (1896–   ); Aristotle Socrates Onassis (1906–75), the shipowner and pioneer of the supertanker, whose second wife was the widow of the US president John F. Kennedy; Joy Adamson (1910–80), the Austrian-born painter and writer, especially remembered for her stories about lions (*Born Free, Elsa,* etc.); and Dr Edwin E. 'Buzz' Aldrin (1930–   ), the

American astronaut who was the second man to walk on the moon (in 1969).

George V died in 1936, saying, 'How is the Empire?' (or, perhaps, 'Bugger Bognor', on being told that he might go to that seaside resort to convalesce).

✌ 21 January

ON this day ... in 1793 the French King Louis XVI was guillotined in the place de la Révolution, the day after he had been sentenced to death for treason; and the Russian revolutionary leader Nikolai Lenin died in 1924.

In 1976, two Concordes made the first scheduled supersonic flights, with the British Airways Flight 300 from Heathrow to Bahrain taking off almost simultaneously with the Air France Flight 085 from the Charles de Gaulle Airport to Rio de Janeiro; and in 1954 the first nuclear-powered submarine, the US *Nautilus*, was launched on the River Thames.

Born on this day were: Thomas Jonathan 'Stonewall' Jackson (1824–63), the American Confederate general; John Moses Browning (1855–1926), the American inventor of the automatic pistol and machine-gun named after him; Christian Dior (1905–57), the French dress-designer and founder of the international fashion house named after him; the British actor Paul Scofield (1922–   ); the British comedian Benny Hill (1925–92); and the Spanish tenor and conductor Placido Domingo (1934–   ).

✌ 22 January

IN 1879, at the battle of Rorke's Drift, two British officers and eighty men of the 24th regiment fought off attacks of more than 4,000 Zulu warriors and eleven Victoria Crosses were won.

In 1924 the first British Labour government came to power, with Ramsay MacDonald as prime minister.

Born on this day were: Francis Bacon, Baron Verulam (1561–1626), the English philosopher and statesman and (according to Alexander Pope) 'the wisest, brightest, meanest of mankind'; the English poet George Gordon Byron, Baron Byron of Rochdale (1788–1824), who (according to Max Beerbohm) 'would be all forgotten today if he had lived to be a florid old gentleman'; the Swedish dramatist, novelist, and painter August Strindberg (1849–1912); the American film director David Lewelyn Wark Griffith (1875–1948); and the Russian-born choreographer and ballet director George Balanchine (1904–83).

✌ 23 January

ON this day . . . in 1806 William Pitt 'the younger' (the youngest British prime minister) died at the age of 46, saying (according to various accounts): 'Oh, my country! how I love my country!' *Or*, 'Oh, my country! how I leave my country!' *Or*, 'My country! oh, my country!' *Or*, 'I think I could eat one of Bellamy's veal pies.'

Born on this day were: the Italian composer and pianist and piano manufacturer Muzio Clementi (1752–1832); the French painter Édouard Manet (1832–83), who was one of the forerunners of the Impressionist movement; the English composer Rutland Boughton (1878–1960); the Russian-born film producer Sergei Mikhailovich Eisenstein (1898–1948); and the Belgian-born jazz guitar virtuoso Django Reinhardt (1910–53).

✌ 24 January

ON this day . . . in 1915 the German cruiser *Blücher* was sunk by a British patrolling squadron commanded by Vice-Admiral Sir David Beatty at the battle of Dogger Bank.

In 1965 Sir Winston Churchill (who was First Lord of the Admiralty at the time of the battle of Dogger Bank) died at the age of 90. (He had predicted that he would die on the same day of the year as his father, Lord Randolph Churchill, who did in fact die on this day in 1895.)

Born on this day were: the English dramatist and poet William Congreve (1670–1729); Frederick II of Prussia, 'Frederick the Great' (1712–86), who was a grandson of George I of Great Britain; Pierre Augustin Caron de Beaumarchais (1732–99), the French playwright; the English Liberal statesman Charles James Fox (1749–1806), described by the historian Sir George Macaulay Trevelyan as 'Our first great states-man of the modern school'; the German writer and composer Ernst Theodor Wilhelm Hoffmann (1776–1822), whose *Tales* inspired Offen-bach's *Tales of Hoffmann*; the American novelist Edith Wharton (1862–1937); and the Austrian-born novelist Vicki Baum (1888–1960).

℘ 25 January

IN 1900, on this second and final day of the battle of Spion Kop (one of the earlier engagements of the Boer War), the British forces suffered severe losses: and the young Winston Churchill was present as a war correspondent.

Born on this day were: Edmund Campion (1540–81), the first of the Jesuit martyrs; the Hon. Robert Boyle (1627–91), one of the founders of modern chemistry and physics; Scotland's national poet Robert Burns (1759–96), whose birthday has long been celebrated on this date as 'Burns Night' by Scotsmen world-wide; Hugh Cecil Lowther, who as Lord Lonsdale founded and gave his name to the Lonsdale Belt for box-ing; the English novelist and short-story writer W(illiam) Somerset Maugham (1874–1965); the English novelist Virginia Woolf (1882–1941), who was a leading member of the Bloomsbury Group of intellectuals; the English actress and centenarian Gwen Ffrangcon-Davies (1891–1992); Paul Henri Spaak (1899–1972), the Belgian prime minister who was the first president of the UN General Assembly and a founding father of the EEC; and Eduard Shevardnadze (1928–   ), the former Soviet minister of foreign affairs.

℘ 26 January

ON this day . . . in 1788 Captain Arthur Phillip landed at Sydney Cove with a fleet of ships carrying convicts from England; and the day has since been observed as Australia's national day. (Captain Phillip eventually became the first governor of New South Wales.)

In 1885 General Gordon was murdered on the palace steps at Khartoum, at the end of a siege of ten months, and two days before the arrival of a relief expedition.

Born on this day were: General Douglas MacArthur (1880–1964), the American commander of the Allied powers who accepted the surrender of Japan in 1945; Sir Henry Cotton (1907–87), the British golf champion who was knighted posthumously; the French jazz violinist Stéphane Grappelli (1908–), the founder with the guitarist Django Reinhardt of the famous Hot Club de France, whose career has spanned seven decades; and Jacqueline du Pré (1945–87), the English cellist whose playing career ended in 1973 through multiple sclerosis.

✃ 27 January

WOLFGANG Amadeus Mozart (1756–91), one of the greatest musical prodigies, was born on this day in Salzburg.

Also born on this day were: Henry Greathead (1757–1816), the Englishman who invented the first purpose-built lifeboat (in 1790); Charles Lutwidge Dodgson (1832–98), the English mathematician who used the pseudonym 'Lewis Carroll' in writing *Alice in Wonderland*; Wilhelm II (1859–1941), the third German emperor (known disparagingly in World War I as 'Kaiser Bill'), who was forced to abdicate two days before the Armistice in 1918; and the American song composer Jerome Kern (1885–1945).

In 1926, John Logie Baird gave the first public demonstration of television to members of the Royal Institution.

✃ 28 January

ON this day . . . in 1547 Henry VIII died (exactly 100 years after the birth of his father Henry VII) and was succeeded by his only son Edward VI; in 1596 Sir Francis Drake died (of dysentery) aboard his ship, off Porto Bello ('Take my drum to England, hang et by the shore, | Strike et when your powder's runnin' low; | If the Dons sight Devon, I'll quit the port o'Heaven, | An' drum them up the Channel as we drummed them long ago'—Sir Henry Newbolt); and the Canadian army surgeon John McCrae was killed in action in 1918. (His poem 'In Flanders Fields' inspired 'Poppy Day': '. . . If ye break faith with us who die I We shall not sleep, though poppies grow I In Flanders fields.')

Born on this day were: John Baskerville (1706–75), the English printer who gave his name to the 1763 edition of the Bible; the French novelist and dancer known as 'Colette' (Sidonie Gabrielle Claudine Colette, 1873–1954); Auguste Piccard (1884–1962), the Swiss scientist who at various times was the joint holder of altitude records (in balloons) and the record for ocean descent; and the Polish-born pianist Artur Rubinstein (1887–1982), who was made an honorary KBE at the age of 90.

✌ 29 January

ON this day . . . in 1856 the Victoria Cross, the medal awarded to British and Commonwealth armed forces for outstanding bravery 'on the field of battle', was instituted by Queen Victoria. (The medal was at first made from the metal of cannon captured from the Russians at Sevastopol, until the supply came to an end in 1942.)

Born on this day were: Thomas Paine (1737–1809), the English political philosopher whose treatise *The Rights of Man* forced him into exile in France and America; John Callcott Horsley (1817–1903), the English artist who designed the first commercial Christmas cards, in 1843; William McKinley (1843–1901), the US president who was assassinated in the first year of his second term of office; the Russian playwright and short-story writer Anton Chekhov (1860–1904); the English composer Frederick Delius (1862–1934); the Spanish novelist Vicente Blasco Ibáñez (1867–1928), best remembered for *The Four Horsemen of the Apocalypse*; and the English composer Havergal Brian (1876–1972), whose prolific output included twenty-seven symphonies, four operas, and a number of other major works written after his 70th birthday.

✌ 30 January

ON this day ... in 1649 Charles I was beheaded at Whitehall by the hangman Richard Brandon.

In 1948 the Indian leader Mahatma Gandhi was shot dead by a Hindu fanatic in Delhi.

Born on this day were: Walter Savage Landor (1775–1864), the English poet who was caricatured as 'Boythorn' in Dickens's *Bleak House*; Sir William Jenner (1815–98), the Physician in Ordinary to Queen Victoria who discovered the difference between typhus and typhoid fever; Franklin Delano Roosevelt (1882–1945), the thirty-second US president, who was a distant cousin of President Theodore Roosevelt; and the English actress Vanessa Redgrave (1937–   ), a daughter of the actor Sir Michael Redgrave.

In 1933 Hitler was named as chancellor by President Hindenburg.

✌ 31 January

ON this day ... in 1606 Guy Fawkes was hanged, drawn, and quartered for his part in the Gunpowder Plot of the previous November; and in 1788 Charles Edward Stuart ('Bonnie Prince Charlie'), the great-great-grandson of the king whom Guy Fawkes had intended to blow up, died in Rome, having assumed the title of Charles III of Great Britain.

Born on this day were: Franz Schubert (1797–1828), the Austrian composer who died in his 32nd year with his Symphony No. 8 in B minor 'Unfinished'; the Russian ballerina Anna Pavlova (1882–1931); the American film actor and comedian Eddie Cantor (1892–1964); the American novelist Norman Mailer (1923–   ); and Beatrix, queen of The Netherlands (1938–   ).

In 1943 Field Marshal Paulus surrendered the German 6th Army to the Russians at Stalingrad, in disobedience to the orders of Hitler.

FEBRUARY &

BORN on this day were: Dame Clara Butt (1872–1936), the English contralto and the first English musician to receive the DBE, the feminine equivalent of a knighthood (in 1920); and Sir Stanley Matthews (1915–   ), the first English professional footballer to be knighted while still playing professionally (in 1965, at the age of 50).

Also born on this day were: Sir Edward Coke (1552–1634), the English judge who prosecuted Sir Walter Ralegh and the 'Gunpowder Plot' conspirators (and maintained that 'a man's house is his castle'); the English actor John Philip Kemble (1757–1823), elder brother of Sarah Siddons; the Irish-American composer and conductor Victor Herbert (1859–1924); the Irish-American film director John Ford (1895–1973); the English writer Stephen Potter (1900–69), who coined the word 'Gamesmanship' (*The Art of Winning Games Without Actually Cheating*); Alan Strode Campbell Ross (1907–   ), the English university professor who coined the terms 'U' and 'non-U'; and Boris Yeltsin (1931–   ), the first president of Russia to be freely elected.

The first volume of the *Oxford English Dictionary*, *A–Ant*, was published in 1884.

&a 2 *February*

BORN on this day were two of the greatest violinists of modern times: the Austrian-born Fritz Kreisler (1875–1962) and the Russian-born Jascha Heifetz (1901–87), who both became American citizens.

Also born on this day were: Nell Gwyn (1650–87), the English actress and mistress of Charles II; the Irish writer James Joyce (1882–1941), who spent the larger part of his life in France; and the former president of France, Valéry Giscard d'Estaing (1926–   ).

In 1943 the battle of Stalingrad ended with the final surrender of the German forces.

❧ 3 *February*

THE Marquess of Salisbury (1830–1903), who was three times prime minister between 1885 and 1902, including the years of the South African War, was born on this day; and in 1960, in South Africa, another British prime minister, Harold Macmillan, made his historic speech to the Parliament in Cape Town. ('The wind of change is blowing through the continent.')

Also born on this day were: the German composer Felix Mendelssohn (1809–47), whose family added 'Bartholdy' to their name when they converted from Judaism to Christianity in 1816; Elizabeth Blackwell (1821–1910), the first Englishwoman to be registered as a doctor; Walter Bagehot (1826–77), the English political economist and journalist known as 'the Spare Chancellor'; Hugo Junkers (1859–1935), the German aircraft engineer and designer (in 1915) of the first all-metal aeroplane; Lord Trenchard (1873–1956), who became known as 'the father of the Royal Air Force'; and James Michener (1907–   ), the author of *Tales of the South Pacific*, which won the 1947 Pulitzer prize for fiction and was adapted for the successful musical film *South Pacific*.

In 1911 Robert Tressell, the author of *The Ragged Trousered Philanthropists*, died in a workhouse in Liverpool, with his now famous work unpublished.

❧ 4 *February*

ON this day . . . in 1945 the Yalta Conference began in the USSR, with the 'big three', Churchill, Roosevelt, and Stalin, getting together to arrange the fate of the world.

Born on this day were: Marshal Voroshilov (1881–1969), the commander of the Russian forces at the siege of Leningrad and later president of the USSR 1953–60; the French painter Fernand Léger (1881–1955); Charles Lindbergh (1902–74), the American aviation pioneer; Lord Shawcross (1902–   ), who as Sir Hartley Shawcross was Britain's chief prosecutor at the international trials of the leading Nazi war criminals at Nuremberg; and Dietrich Bonhoeffer (1906–45), the German Lutheran pastor and theologian who was hanged by the Nazis for his alleged involvement in the attempted assassination of Hitler.

৯৯ 5 February

BORN on this day were: John Boyd Dunlop (1840–1921), the Scottish inventor who gave his name to a pneumatic tyre; and the American-born inventor Sir Hiram Stevens Maxim (1840–1916), who gave his name to the first fully automatic machine-gun.

Also born on this day were: Sir Robert Peel (1788–1850), the British prime minister who gave both his first name and his surname to nicknames for British policemen, 'bobby' and 'peeler'; the American evangelist Dwight Lyman Moody (1837–99), invariably associated with his musical colleague Ira D. Sankey; the British playwright Frederick Lonsdale (1881–1954); Captain W(illiam) E(arl) Johns (1893–1968), the World War I pilot and author who created 'Biggles'; the American author William Burroughs (1914–   ); and the British author and broadcaster Frank Muir (1920–   ).

৯৯ 6 February

ON this day . . . in 1958, in what became known as 'the Munich air disaster', twenty-three passengers out of forty-four on a British aircraft were killed, among them a number of Manchester United and England footballers and some sportswriters.

George Herman ('Babe') Ruth (1895–1948), the American baseball player who held the record for the most appearances in the World Series matches and was the first to score sixty home runs in a season, Billy Wright (1924–   ), the first England footballer to win 100 international caps, and Freddie Trueman (1931–   ), the first England cricketer to take 300 wickets in Test cricket, were born on this day.

Also born on this day were: Queen Anne (1665–1714), the second daughter of James II and the last of the Stuart sovereigns; the English actor Sir Henry Irving (1838–1905); the English actor Russell Thorndike (1885–1972), the brother of Dame Sybil Thorndike and author of the

'Doctor Syn' stories; the Chilean concert pianist Claudio Arrau (1903–91); Ronald Reagan (1911–   ), the fortieth US president and the oldest (in 1980) to be elected to the office; the English writer and broadcaster Denis Norden (1922–   ); and the English writer Keith Waterhouse (1929–   ).

℘ 7 February

BORN on this day were: the English novelist Charles (John Huffham) Dickens (1812–70); and the American novelist (Harry) Sinclair Lewis (1885–1951), the first American recipient of the Nobel prize for literature, though he had earlier refused the Pulitzer prize.

Also born on this day were: Sir Thomas More (1478–1535), the first layman to become Lord Chancellor of England, who was executed for high treason for refusing to recognize Henry VIII as head of the Church (and is said to have drawn his beard aside as he placed his head on the block, saying, 'This hath not offended the king.'); and the American pianist and songwriter James Hubert 'Eubie' Blake (1883–1983), best remembered for his compositions 'Memories of You' and 'I'm Just Wild About Harry', who said at his 100th birthday celebrations, 'If I'd known how long I was going to live, I'd have taken more care of myself.'

In 1882 the American pugilist John L. Sullivan became the last of the bare-knuckle world heavyweight champions with his defeat of Paddy Ryan at Mississippi City.

℘ 8 February

ON this day . . . in 1587 Mary, Queen of Scots, was beheaded in Fotheringhay Castle (in the English county of Northamptonshire) for her alleged part in a conspiracy to usurp Elizabeth I. (Her son, James VI of Scotland and later James I of England, who had succeeded Elizabeth in 1603, ordered her remains to be removed to Westminster in 1612.)

Born on this day were: John Ruskin (1819–1900), the English art critic and philanthropist; William Tecumseh Sherman (1820–91), the American Civil War general ('There is many a boy . . . who looks on war as all glory, but boys, it is all hell.'); Jules Verne (1828–1905), the French novelist who was one of the first writers of science fiction; Lord Brabazon (1884–1964), the first Englishman to fly in England; and Billy Bishop (1894–1956), the Canadian air ace of World War I.

In 1904 the Russo-Japanese War began.

℘ 9 *February*

ON this day ... in 1567 Lord Darnley, the second husband of Mary, Queen of Scots, was murdered in his sick-bed in a house near Edinburgh. (The house was blown up by gunpowder and Mary and the Earl of Bothwell, who became her third husband some three months later, were believed to have been involved in the murder.)

Born on this day were: William Henry Harrison (1773–1841), the first US president to die in office, a month after his inauguration; Sir Francis Pettit Smith (1808–74), the English inventor of the first screw-propelled ships; Lydia E. Pinkham (1819–83), the American feminist and inventor of a 'vegetable compound' immortalized in song; Sir Anthony Hope (1863–1933), the English novelist who invented the land of 'Ruritania' for his novel *The Prisoner of Zenda*; the British actress Mrs Patrick Campbell (1865–1940); the Austrian composer Alban Berg (1885–1935); the England cricketer Jim Laker (1922–86), who took a record nineteen wickets in a Test match versus Australia in 1956; and the Irish playwright Brendan Behan (1923–64).

In 1916 conscription began in Great Britain as the Military Service Act became effective. ('The unheroic dead who fed the guns . . . Those doomed, conscripted, unvictorious ones'—Siegfried Sassoon.)

℘ 10 *February*

BORIS Pasternak (1890–1960), the Russian poet and novelist whose *Dr Zhivago* gave him the 1958 Nobel prize for literature (which he declined for ideological reasons), was born on this day; and in 1837 the Russian poet Alexander Pushkin died from wounds received in a duel.

Also born on this day were: Charles Lamb (1775–1834), the English essayist and poet who used the pseudonym 'Elia'; Samuel Plimsoll (1824–98), the English social reformer who gave his name to the compulsory load-line for ships and the rubber-soled canvas shoe; the English novelist Howard Spring (1889–1965); Harold Macmillan (1894–1986), the British prime minister 1957–63, and later the Earl of Stockton; the German playwright and poet Bertolt Brecht (1898–1956); Joyce Grenfell (1910–79), the English actress and monologuist; Larry Adler (1914–   ), the American mouth-organist; and the American swimmer Mark Spitz (1950–   ), who was the first to win more than five gold medals in one Olympiad.

℘ 11 February

BORN on this day were: William Henry Fox Talbot (1800–77), the English pioneer of photography who took the earliest surviving photograph, in 1835; and Thomas Alva Edison (1847–1931), the American inventor who was the first man to record sound, in 1877.

Also born on this day were: the French writer Bernard le Bovier de Fontenelle (1657–1757), who died a month short of his 100th birthday, having explored almost all forms of literature; King Farouk I (1920–65), the last king of Egypt, who abdicated in 1952; and Sir Vivian Fuchs (1908–   ), the geologist and explorer who, in 1958, became the first man to traverse the Antarctic.

In 1858 Bernadette Soubirous, a French miller's daughter (baptized Marie Bernarde), is said to have seen the apparition of the Virgin Mary at Lourdes. (She was canonized as St Bernadette in 1933.)

In 1975 Mrs Margaret Thatcher became the first woman to lead the British Conservative Party.

℘ 12 February

ON this day ... in 1554 Lady Jane Grey, the queen of England for thirteen days, following the death of Edward VI on 6 July 1553, was beheaded on Tower Hill with her husband Lord Guildford Dudley. (She was barely 17 years old.)

Born on this day were: Jan Swammerdam (1637–80), the Dutch naturalist who discovered red corpuscles; the English naturalist Charles Darwin (1809–82); Abraham Lincoln (1809–65), the sixteenth US president; John Graham Chambers (1843–83), the English athlete who drew up the Queensberry Rules for boxing, under the supervision of the Marquis of Queensberry; Marie Lloyd (1870–1922), the English music-hall artiste; and the English playwright and novelist R(onald) F(rederick) Delderfield (1912–72).

In 1709 Alexander Selkirk, the Scottish seaman whose adventures inspired the creation of Daniel Defoe's *Robinson Crusoe*, was taken off Juan Fernandez Island after more than four years of living there alone.

℀ 13 *February*

ON this day . . . in 1542 Catherine Howard, the fifth wife of Henry VIII, was beheaded on conviction of adultery.

In 1689 William III and Mary II were proclaimed king and queen of England, following the interregnum. (James II had abdicated in 1688.)

Born on this day were: John Hunter (1728–93), the Scottish surgeon and pioneer of medicine; Lord Randolph Churchill (1849–95), the Conservative statesman and father of Sir Winston Churchill; the Russian bass singer Feodor Ivanovich Shalyapin (1873–1938), also known as 'Chaliapin'; Georges Simenon (1903–89), the Belgian novelist who created 'Inspector Maigret'; and Dame Helen Gardner (1908–86), who edited *The New Oxford Book of English Verse*, 1972.

℀ 14 *February*

ON this day . . . during the latter half of the third century a Roman priest and a bishop of Terni were put to death within a few years of each other. (They were both celebrated as St Valentine; and the rural tradition of birds choosing their mates on this day and young lovers finding their valentine would seem to have come from this coincidence of dates.)

In 1929 seven men, most of them members of the Moran gang, were shot down in a Chicago garage by rival gangsters disguised as policemen. (This episode has since become known as 'the St Valentine's Day Massacre'.)

In 1400 the deposed Richard II (it is generally believed) was murdered in Pontefract Castle, Yorkshire; in 1779 Captain James Cook was murdered by natives of the Hawaiian Islands, the year after he had discovered the islands; and in 1797 the Spanish fleet was defeated by the British under Admiral Jervis (with Nelson in support) at the battle of Cape St Vincent, off Portugal.

Born on this day were: Nicolas Copernicus (1473–1543), the Polish astronomer who originated and gave his name to the system stating that the earth revolves about the sun; Latham Sholes (1819–90), the American inventor of the first modern keyboard typewriter; Quintin Hogg (1845–1903), the founder of the polytechnic in London, whose son and grandson both became Lord Chancellor; and the American comedian and actor Jack Benny (1894–1974).

ℬ 15 *February*

GALILEO Galilei (1564–1642), the Italian astronomer and physicist, was born on this day in Pisa, where he demonstrated his theorem of equal velocity from the famous 'leaning tower'.

In 1942 Singapore was surrendered to the Japanese: and Churchill described the defeat as 'the worst disaster and largest capitulation in British history'.

Also born on this day were: Jeremy Bentham (1748–1832), the English philosopher and social reformer who introduced the registration of births, deaths, and marriages; Sir William Henry Preece (1834–1913), the British electrical engineer who introduced the first telephone receivers; Sir Ernest Shackleton (1874–1922), the British explorer who died on his fourth Antarctic expedition; Lawrence Wright, alias Horatio Nicholls (1888–1964), the English songwriter and music publisher who was known as 'the father of Tin Pan Alley'; and the American songwriter Hyam Arluck, alias Harold Arlen (1905–86).

In 1971 decimal coinage was adopted in Great Britain.

ℬ 16 February

BORN on this day were: Henry Brookes Adams (1838–1918), the American historian whose grandfather and great-grandfather were US presidents; and George Macaulay Trevelyan (1876–1962), the English historian whose great-uncle was the historian Lord Macaulay.

In 1940, in a historic episode of World War II, 299 British seamen were rescued from a German prison ship, the *Altmark*, in a Norwegian fiord. (The armed party from the destroyer HMS *Cossack* announced their arrival with the now famous cry, 'The Navy's here!')

Also born on this day were: Philip Melanchthon (1497–1560), the German Protestant reformer and associate of Luther; the French soldier and Protestant leader Gaspard de Coligny (1519–72), who was one of the first victims of the Massacre of St Bartholomew; Ernst Heinrich Haeckel (1834–1919), the German naturalist and proponent of Darwinism; Charles Taze Russell (1852–1916), the American preacher and founder of the International Bible Students' Association, whose members were known as Russellites and, after Russell's death, Jehovah's Witnesses; the Welsh operatic baritone Sir Geraint Evans (1922–92); and the American tennis player John McEnroe (1959–   ).

ℬ 17 February

ON this day... in 1864 a Confederate submarine *H. L. Hunley* carried out the first effective attack by a submersible vessel when it rammed a Federal corvette *Housatonic* in Charleston Harbor. (Both vessels were blown up, however, and all the submarine's crew perished.)

Born on this day were: the Swiss scientist Horace Bénédict de Saussure (1740–99), who coined the word 'geology'; the English economist Thomas Robert Malthus (1766–1834), whose *Essay on the Principle of Population* gave rise to the word 'Malthusian'; René Laënnec (1781–1826), the French physician who invented the stethoscope; the English composer Sir Edward German (1862–1936); Andrew Barton Paterson (1864–1941), the Australian journalist and poet who wrote the words of 'Waltzing Matilda'; André Maginot (1877–1932), the French minister of war who initiated and gave his name to the system of fortifications intended to defend France from a German invasion; and the American contralto Marion Anderson (1902–   ), described by Toscanini as 'the voice that comes once in 100 years'.

✌ 18 February

On this day... in 1478 George, the Duke of Clarence, who had opposed his brother Edward IV, was murdered in the Tower of London (according to various accounts by being drowned in a butt of malmsey).

Mary I (1516–58), the daughter of Henry VIII by his first wife Catherine of Aragon, was born on this day. (During her reign, from 1553 until her death, her persecution of Protestants gave rise to her nickname of 'Bloody Mary'.) Martin Luther, the German Protestant reformer, died in 1546. ('If I had heard that as many devils would set on me in Worms as there were tiles on the roofs, I should none the less have ridden there.')

Also born on this day were: Count Alessandro Volta (1745–1827), the Italian scientist who invented the electric battery and gave his name to the unit of electromotive force, the volt; Marshall Hall (1790–1857), the English physiologist who discovered the reflex action; and the English novelist (and chronicler of World War II) Len Deighton (1929–   ).

✌ 19 February

On this day... in 1915 British and French warships began the attack on the Turkish forts at the mouth of the Dardanelles, in the abortive expedition to force the straits of Gallipoli.

Born on this day were: the English actor and dramatist David Garrick (1717–79), who also wrote the words of the song 'Heart of Oak'; and the English actor Sir Cedric Hardwicke (1893–1964).

Also born on this day were: Luigi Boccherini (1743–1805), the Italian composer who was labelled as 'the wife of Haydn', from his emulation of Haydn's works; the Italian soprano Adelina Patti (1843–1919); the American bandleader and composer Stan Kenton (1912–79); and Prince Andrew, the Duke of York (1960–   ), the second son of Queen Elizabeth and Prince Philip.

℀ 20 *February*

ON this day ... in 1962 Colonel John Glenn became the first American astronaut and the first man to orbit the earth more than once, making three orbits.

Born on this day were: William Prescott (1726–95), the American revolutionary leader and hero of the battle of Bunker Hill (in 1775); Henry James Pye (1745–1813), the English Poet Laureate 1790–1813, whose 'poetry' was the butt of his contemporaries; Adam Black (1784–1874), the Scottish publisher and founder of A. & C. Black, the publishers of the original *Who's Who*; and Karl Czerny (1791–1857), the Austrian pianist, composer, and piano teacher who was a pupil of Beethoven and a teacher of Liszt.

℀ 21 *February*

ON this day ... in 1916 the battle of Verdun began with an unprecedented bombardment by the German artillery and continued into the summer, with great losses to both French and German forces. (The breakthrough was not achieved, however: 'Ils ne passeront pas'—Marshal Pétain.)

Born on this day were: Cardinal John Henry Newman (1801–90), the English theologian who was converted to the Catholic faith in 1845 and created a cardinal in 1879; the French composer Léo Delibes (1836–91), best remembered for his ballet *Coppélia*; George Lansbury (1859–1940), the British Labour politician who founded the *Daily Herald*, in 1912; Andrés Segovia (1893–1987), the Spanish guitarist who pioneered classical guitar techniques; the English poet W(ystan) H(ugh) Auden (1907–73); the British airman Sir Douglas Bader (1910–82), who after losing both legs in a flying accident in 1931 became one of the most successful fighter pilots of World War II; and Robert Mugabe (1924–), the first prime minister and president of Zimbabwe.

In 1595 the English poet and Jesuit martyr Robert Southwell was hanged, drawn, and quartered at Tyburn, after three years of imprisonment and torture.

✌ 22 February

ON this day... in 1797 the 'last invasion of Britain' took place when some 1,400 Frenchmen landed at Fishguard, in Wales.

Born on this day were: George Washington (1732–99), the American soldier and the first president of the USA; the German philosopher Arthur Schopenhauer (1788–1860); the German scientist Heinrich Hertz (1857–94), who gave his name to the units of frequency (the kilohertz, etc.); Lord Baden-Powell (1857–1941), the British soldier who (in 1908) founded the Boy Scouts; his wife Olave, Lady Baden-Powell (1889–1977), the World Chief Guide from 1930; and the British actor Sir John Mills (1908–   ), whose wife is the playwright Mary Hayley Bell.

In 1879 the first chain store was originated with the opening of Frank Winfield Woolworth's 'nothing over five cents' shop at Utica, New York.

✌ 23 February

BORN on this day were: Samuel Pepys (1633–1703), the English naval administrator whose famous diary opened on 1 January 1660; George Frideric Handel (1685–1759), the German-born English composer; and George Frederick Watts (1817–1904), the English painter.

The English poet John Keats died in Rome in 1821, aged 25. ('When I have fears that I may cease to be | Before my pen has glean'd my teeming brain. | ... Then on the shore | Of the wide world I stand alone, and think | Till love and fame to nothingness do sink.')

In 1885 John Lee ('the man they couldn't hang') survived three attempts to hang him at Exeter Prison, as the trap failed to open. (He had been sentenced to death for murdering his employer, but he was eventually released in 1917 and emigrated to America, where he married and lived until his death in 1933.)

In 1946 Tomoyuki Yamashita, the Japanese general who commanded the Philippines campaign in World War II, was hanged for war crimes, in Manila.

24 *February*

In 1525, in the first of the Franco-Habsburg Wars, the Holy Roman Emperor Charles V captured the French king Francis I at the battle of Pavia, in Italy.

Born on this day were: Don John of Austria (1545–78), the commander of the combined fleets which defeated the Turks at the battle of Lepanto in 1571; Samuel Wesley (1766–1837), the organist and composer and younger son of Charles Wesley, the Methodist hymn-writer; Wilhelm Grimm (1786–1859), the author (with his elder brother Jacob) of the famous *Fairy Tales*; the Irish novelist George Moore (1852–1933); Arnold Dolmetsch (1858–1940), the French-born English musical instrument maker and restorer of early instruments; Brian Close (1931–), the youngest cricketer to have played for England (in 1949); and Denis Law (1940–), the youngest footballer to have played for Scotland (in 1958).

25 *February*

On this day . . . in 1601 Robert Devereux, the second Earl of Essex and a former favourite of Elizabeth I, was beheaded in the Tower of London for high treason.

Sir Christopher Wren died in 1723, aged 90. (He was buried in St Paul's Cathedral, where an inscription written by his son over the interior of the north door reads: 'Si monumentum requiris, circumspice.' (If you would seek his monument, look around.))

Born on this day were: the French Impressionist painter Pierre Auguste Renoir (1841–1919); John Foster Dulles (1888–1959), the US secretary of state 1953–9, whose statement (in 1956), 'You have to take chances for peace just as you have to take chances in war . . . If you are scared to go to the brink you are lost', gave rise to the word 'brinkmanship'; Dame Myra Hess (1890–1965), the English concert pianist (who also died on this day); Victor Sylvester (1900–78), the English ballroom dancing champion and bandleader whose name became synonymous with strict tempo dance music; John Arlott (1914–91), the English writer and broadcaster whose name became synonymous with the game of cricket; and the English writer and composer Anthony Burgess (1917–).

✑ 26 February

ON this day . . . in 1839 the first Grand National Steeplechase (though known until 1847 as the Grand Liverpool Steeplechase) was won by Jem Mason on Lottery.

Born on this day were: the French writer Victor Hugo (1802–85); William Frederick Cody (1846–1917), the American frontiersman and showman known as 'Buffalo Bill', who gave his name to the town of Cody in Wyoming; Émile Coué (1857–1926), the French pharmacist who gave his name to a method of psychotherapy by auto-suggestion, Couéism ('Every day, in every way, I am getting better and better . . .'); Frank Bridge (1879–1941), the English composer and mentor of Benjamin Britten; and Orde Charles Wingate (1903–44), the British general who led the 'Chindits', the specially trained jungle fighters in Burma in World War II.

✑ 27 February

ON this day . . . in 1933 the burning down of the Reichstag building in Berlin gave the Nazis the opportunity to suspend personal liberty with increased powers.

Born on this day were: the American poet Henry Wadsworth Longfellow (1807–82); Dame Ellen Terry (1847–1928), the first English actress to receive the GBE; Sir Hubert Parry (1848–1918), the English composer who is best remembered for his setting of Blake's poem 'Jerusalem'; Sir Cecil Spring-Rice (1859–1918), the British ambassador to the USA 1912–18, who is best remembered for his poem 'I vow to thee, my country', which he wrote a few weeks before his death and his final return to England; the Italian tenor Enrico Caruso (1873–1921); Charles Herbert Best (1899–1978), the Canadian physiologist who with Sir Frederick Grant Banting discovered insulin; the American novelist John Steinbeck (1902–68), the recipient of the Nobel prize for literature in 1962; the English novelist Lawrence Durrell (1912–90); and the English-born actress Elizabeth Taylor (1932–   ).

John Evelyn died in 1706, though his famous diary was not published until 1818.

℘ 28 *February*

ON this day . . . in 1900 General Buller led the British force in ending the siege of Ladysmith, in Natal, which had lasted for some four months.

Born on this day were: Michel Eyquem de Montaigne (1533–92), the French essayist and moralist ('I have only made up a bunch of other people's flowers, and . . . provided the string that ties them together.'); the French scientist René Antoine de Réaumur (1683–1757), the inventor of the thermometer named after him; the French rope-artist known as Charles Blondin (1824–97), who was the first man to cross Niagara Falls on a tightrope, in 1859; Philip Showalter Hench (1896–1965), the American scientist who discovered cortisone and was joint recipient of the 1950 Nobel prize for physiology and medicine; Linus Carl Pauling (1901–   ), the American scientist who was the first to win two full Nobel prizes (the 1954 chemistry prize and the 1962 peace prize); and the English poet and critic Sir Stephen Spender (1909–   ).

℘ 29 *February*

AN additional day of the year in February in every fourth year, or leap year, was first added to the other 365 days with the introduction of the Julian calendar in 46 BC. The Gregorian (or New Style) calendar modified this by excluding the leap years at the turn of the centuries not divisible by 400. (The years 1700, 1800, and 1900 were accordingly not leap years, whereas the year 2000 will include this day.)

St Hilarius, the forty-sixth pope, died in 468; and St Oswald, archbishop of York, died in 992. (Their feast-days are celebrated on 28 February.)

In 1528 the Scottish martyr Patrick Hamilton was burnt at the stake for heresy. (It is said that his death did more for the Scottish Reformation than the continuation of his life could have done.)

Born on this day were: the Marquis de Montcalm (1712–59), the French commander at the battle of Quebec; Ann Lee (1736–84), the English religious mystic who founded the 'Shakers' sect in the USA, in 1776; the Italian operatic composer Gioacchino Rossini (1792–1868); John

Philip Holland (1840–1914), the Irish inventor who designed the first submarines for the US Navy (in 1900) and the Royal Navy (in 1901); the American bandleader Jimmy Dorsey (1904–57); and Mario Andretti (1940–   ), the American Grand Prix driver and World champion in 1978.

MARCH ✑

✑ 1 March

THIS day is the festival of St David (died c.600), the patron saint of Wales; and born on (and named from) St David's Day were: the British film actor David Niven (1910–83); the British Olympic show-jumper and professional world champion David Broome (1940–   ); and David Scott Cowper (1942–   ), the British record-breaking yachtsman and solo circumnavigator.

Also born on this day were: Frédéric Chopin (1810–49), the Polish composer and pianist; the English sculptor Sir Thomas Brock (1847–1922); the American sculptor Augustus Saint-Gaudens (1848–1907); Lytton Strachey (1880–1932), the English biographer and one of the Bloomsbury Group; Oskar Kokoschka (1886–1980), the Austrian-born Expressionist painter and writer; and the American bandleader and composer Glenn Miller (1904–44), who became a World War II legend with his US Army Band.

In 1954 the first American hydrogen bomb was 'officially' detonated at the Bikini Atoll, in the Marshall Islands, some seventeen months after the first reported tests.

✑ 2 March

ON this day . . . in 1877 Rutherford Birchard Hayes was declared the nineteenth US president by an electoral commission, following a dispute over the 1876 ballots. (He was the only US president to be elected by this process.)

Born on this day were: Sir Thomas Bodley (1545–1613), the English statesman and bibliophile who founded and gave his name to the Bodleian Library at Oxford; Samuel Houston (1793–1863), the American soldier and statesman who gave his name to the city of Houston, in Texas; the Czech composer Bedřich Smetana (1824–84); the German composer Kurt Weill (1900–50); Mikhail Gorbachev (1931–   ), the

Soviet leader 1985–91, and the last to preside over the USSR; and Dame Naomi James (1949–   ), the New Zealand-born yachtswoman and the first to circumnavigate solo via Cape Horn.

In 1958 the British Commonwealth Trans-Antarctic expedition, led by Dr Vivian Fuchs, completed the first surface crossing of the South Pole.

In 1969 the French-built Concorde made its maiden flight.

℘⃞ 3 March

IN 1943 the battle of the Bismarck Sea ended after three days, with heavy losses sustained by the Japanese ships from Allied air attacks.

Born on this day were: Edward Herbert, Lord Herbert of Cherbury (1583–1648), the English statesman and Royalist poet known as the 'father of Deism'; the Royalist poet Edmund Waller (1606–87); William Godwin (1756–1836), the English political writer and novelist and father-in-law of the poet Shelley; the English actor William Macready (1793–1873); George Mortimore Pullman (1831–97), the American inventor of the railway sleeping-car; Alexander Graham Bell (1847–1922), the Scottish-born American who is credited with the invention of the telephone; Sir Henry Wood (1869–1944), the English conductor who inaugurated the 'Proms'; and Edward Thomas (1878–1917), the English poet who was killed at the battle of Arras.

In 1918 the peace treaty of Brest-Litovsk between Russia and Germany was signed, depriving Russia of the Baltic states, Finland, the Ukraine, and Poland.

℘⃞ 4 March

ON this day ... in 1461 Henry VI was deposed and the Duke of York proclaimed king as Edward IV; and in 1980 Robert Mugabe was invited to form a government in Rhodesia (later Zimbabwe) after his party had won an absolute majority in the elections.

Born on this day were: Antonio Vivaldi (1678–1741), the Italian composer and violinist; Sir Henry Raeburn (1756–1823), the portrait painter known as 'the Scottish Reynolds'; the English novelist and playwright

Alan Sillitoe (1928–   ); and the Dutch conductor Bernard Haitink (1929–   ).

✌ 5 March

In 1946 Winston Churchill made his historic 'Iron Curtain' speech at Fulton, Missouri. ('From Stettin in the Baltic to Trieste in the Adriatic, an Iron Curtain has descended across the Continent.') The Russian dictator Joseph Stalin, the creator of the Iron Curtain or the 'cold war', died in 1953.

Born on this day were: Henry II (1133–89); Gerhardus Mercator (1512–94), the Flemish geographer and map-maker and pioneer of modern cartography; the Italian painter Giovanni Battista Tiepolo (1696–1770); Étienne Jules Marey (1830–1903), the French physiologist and pioneer of animal cinematography and camera design; William Henry Beveridge (later Lord Beveridge), the pioneer of Britain's post-war social insurance scheme (1879–1963); and the English actor Rex Harrison (1908–90).

✌ 6 March

On this day . . . in 1836 the siege of El Alamo ended after thirteen days, as fewer than 200 Texans (including Davy Crockett and Colonel James Bowie) were finally overwhelmed by some 3,000 Mexicans. (The battle has since come to be known as 'the Thermopylae of America'.)

Born on this day were four women of varied achievements: Elizabeth Barrett Browning (1806–61), the English poet who was married to Robert Browning; Rose Fyleman (1877–1957), the English writer of children's books and mainly remembered for her poem 'There are fairies at the bottom of our garden . . .'; Valentina Nikolayeva-Tereshkova (1937–   ), the Russian astronaut who was (in 1963) the first woman to orbit the earth; and the New Zealand operatic soprano Dame Kiri Te Kanawa (1944–   ).

Also born on this day were: the Italian sculptor, painter, and poet Michelangelo Buonarroti (1475–1564); Philip Henry Sheridan (1831–88), the American Civil War general and hero of the battle of Cedar Creek;

35

George du Maurier (1834–96), the French-born artist and novelist and grandfather of the British novelist Daphne du Maurier; and Samuel Franklin Cody (1861–1913), the American-born aviator who was the first to make a powered flight in Britain.

℘ 7 *March*

ON this day . . . in 1876 Patent No. 174,465 was issued for the telephone of Alexander Graham Bell, the first practical telephone; and in 1917 the first gramophone record of a jazz band was released. (This featured the Original Dixieland Jazz Band playing 'The Dixie Jazz Band One Step' and 'Livery Stable Blues'.)

Born on this day were: the Italian poet and novelist Alessandro Manzoni (1785–1873), whose death inspired Verdi's *Requiem*; Sir John Herschel (1792–1871), the English astronomer and son of the astronomer Sir William Herschel; Sir Edwin Landseer (1802–73), the English animal painter who modelled the Trafalgar Square lions; the Dutch painter Piet Mondrian (1872–1944), the leader of the Neo-plasticism movement; the French composer Maurice Ravel (1875–1937); and the British statesman Ernest Bevin (1881–1951).

℘ 8 *March*

ON this day . . . in 1702 William III died and (in the words of the song) 'gracious Anne became our queen' and the Vicar of Bray 'became a Tory'.

Born on this day were: the German composer Carl Philipp Emanuel Bach (1714–88), the second son of J. S. Bach; the Italian composer Ruggiero Leoncavallo (1858–1919), whose best-known opera *I Pagliacci* also (and most unusually) had the libretto written by the composer; Kenneth Grahame (1859–1932), the Scottish author and creator of 'Rat', 'Mole', 'Badger', and 'Toad' in his children's classic *The Wind in the Willows*; and the Scottish author Eric Linklater (1899–1974), who created 'Private Angelo'.

In 1910 John Theodore Cuthbert Moore-Brabazon, later Lord Brabazon of Tara, was granted the Royal Aero Club's first certificate for pilots; and the first woman pilot, Mme de Laroche, was issued with a pilot's licence by the Aéro-Club de France.

✇ 9 March

ON this day . . . in 1950 Timothy John Evans was hanged at Pentonville Prison for the murder of his baby daughter. (The wrongful conviction of Evans was established some three years after his execution with the sentencing to death of the real murderer, John Reginald Christie, who confessed to five other murders, including that of Evans's wife.)

Born on this day were: William Cobbett (1763–1835), the English adventurer and political writer remembered for his *Rural Rides*; Vyacheslav Mikhailovich Molotov (1890–1986), the last of the leading Russian revolutionaries; David Garnett (1892–1981), the English novelist and one of the Bloomsbury Group; the English novelist and poet Victoria 'Vita' Sackville-West (1892–1962); the Scottish soprano Dame Isobel Baillie (1895–1983); the American composer Samuel Barber (1910–81); Yuri Gagarin (1934–68), the Russian cosmonaut who was the first man to orbit the earth; and Bobby Fischer (1943–   ), the first American world chess champion.

✇ 10 March

ON this day . . . in 1876 Alexander Graham Bell made his historic telephone call to his assistant Thomas Watson, at a house in Boston, Massachusetts, on the day after he had patented his invention: 'Mr Watson, come here, I want you!'

Born on this day were: Marcello Malpighi (1628–94), the Italian anatomist whose name (as 'Malpighian') is given to several of the body's structures; the Italian poet Lorenzo da Ponte (1749–1838), who wrote the libretti for a number of Mozart's operas; the English sculptor Edward Hodges Baily (1788–1867), who was the sculptor of the statue of Nelson in

Trafalgar Square; John Bacchus Dykes (1823–76), the English clergyman and composer who wrote scores of our best-known hymn tunes; Wyatt Earp (1848–1929), the legendary marshal of Tombstone, Arizona, and the protagonist at 'the gunfight at OK Corral' (in 1881); Henry Watson Fowler (1858–1933), the English lexicographer who with his brother Frank George Fowler edited the first *Concise Oxford Dictionary* (in 1911) and, after his brother's death, the *Dictionary of Modern English Usage* (in 1926), which still bears his name; the French-born Swiss composer Arthur Honegger (1892–1955); the English soprano Dame Eva Turner (1892–1990); the American jazz cornettist Bix Beiderbecke (1903–31), whose life and tragic early death inspired Dorothy Baker's novel *Young Man with a Horn*; and the English conductor Sir Charles Groves (1915–92).

✿ 11 *March*

On this day . . . in 1702 the *Daily Courant*, the first regular English daily newspaper, was published; and the Australian-born newspaper and television magnate Rupert Murdoch (1931–   ) was born on this day.

Also born on this day were: the Italian poet Torquato Tasso (1544–95); Sir Henry Tate (1819–99), the English sugar merchant who endowed and gave his name to the Tate Gallery, in London; Sir Malcolm Campbell (1885–1949), the first holder of world speed records in motor and speed-boat racing and the first motorist to exceed 300 m.p.h. (in 1935, at the age of 50); Jessie Matthews (1907–72), the British actress and singer; and the British prime minister Harold Wilson, 1964–70, 1974–6, Lord Wilson of Rievaulx (1916–   ).

✿ 12 *March*

On this day . . . in 1917 Russian troops mutinied as the 'February Revolution' began. (According to the Julian calendar, still in use in Russia at that time, the date was 27 February.)

Born on this day were: André Le Nôtre (1613–1700), the French pioneer of landscape gardening who designed the gardens at Versailles and

St James's Park in London; George Berkeley (1685–1753), the Irish bishop and philosopher; Thomas Augustine Arne (1710–78), the English composer best remembered for 'Rule, Britannia', from his masque *Alfred*; John Daniell (1790–1845), the English chemist who invented and gave his name to one of the earliest electric batteries, the Daniell cell; the English chemist Sir William Henry Perkin (1838–1907), who produced the first synthetic dye, mauve (in 1856), and discovered the Perkin reaction for making aromatic unsaturated acids; Gabriele d'Annunzio (1863–1938), the Italian poet, novelist, dramatist, soldier, airman, and politician; and the actress and singer Liza Minnelli (1946–   ), the daughter of Judy Garland and the film director Vincente Minnelli.

℘ 13 *March*

On this day . . . in 1781 the German-born astronomer (Sir) William Herschel discovered the planet Uranus, which he named 'Georgium Sidus', in honour of George III; and the American astronomer Percival Lowell (1855–1916), who predicted the discovery of the planet Pluto and founded the Flagstone Observatory in Arizona, was born on this day.

Also born on this day were: Joseph Priestley (1733–1804), the English philosopher and scientist who is credited with the discovery of oxygen; Hugo Wolf (1860–1903), the Austrian composer credited as second only to his fellow-countryman Franz Schubert as a songwriter; and the English novelist Sir Hugh Walpole (1884–1941).

In 1881 Tsar Alexander II was assassinated when a bomb was thrown at him near his palace.

℘ 14 *March*

In 1757 the British admiral John Byng was executed by a firing squad on board HMS *Monarch* at Plymouth, for neglect of duty at an abortive action off Minorca. (This gave rise to Voltaire's frequently quoted phrase from his satire *Candide*: 'In this country it is thought well to kill an admiral from time to time to encourage the others.')

Born on this day were: Georg Philipp Telemann (1681–1767), one of the most prolific German composers; the Austrian composer Johann Strauss the elder (1804–49), not the composer of *The Blue Danube*, which was written by his son of the same name; Mrs Isabella Mary Beeton (1836–65), the English author whose *Household Management* has made her name synonymous with the cookery book; Maxim Gorky (1868–1936), the Russian author whose name was given to the city which was his birthplace; and the German-born mathematical physicist Albert Einstein (1879–1955), whose theories revolutionized the world of science.

Karl Marx, the German philosopher whose revolutionary theories were not to be put to the test until some thirty years after his death, died in 1883, in London.

৪৯ 15 *March*

THIS day is known as 'the ides of March', from the ancient Roman calendar which gave this word to days that divided the month. Julius Caesar was assassinated on this day in 44 BC; and the soothsayer's forewarning of his doom, 'Beware the ides of March', has become a part of the English language as any warning of threatening danger.

Born on this day were: Andrew Jackson (1767–1845), the seventh US president, known as 'Old Hickory'; William Lamb, Viscount Melbourne (1779–1848), the British prime minister who was married to the novelist Lady Caroline Lamb; the English physician John Snow (1813–58), who discovered that cholera is transmitted by contaminated water and was the first to use ether as an anaesthetic; the German scientist Emil von Behring (1854–1917), the first recipient of the Nobel prize in medicine (in 1901); and Leslie Stuart (1866–1928), the English song composer best remembered for 'Soldiers of the Queen' and 'Lily of Laguna'.

In 1949, almost four years after the end of World War II, clothes rationing in Great Britain was officially ended.

✍ 16 *March*

ON this day . . . in 1970 the complete New English Bible was first published.

Born on this day were: James Madison (1751–1836), the fourth US president; Georg Simon Ohm (1787–1854), the German scientist who gave his name to the unit of electrical resistance; William Henry Monk (1823–89), the English organist and the first editor of *Hymns Ancient and Modern*, for which he composed a number of well-known hymn tunes, such as 'Eventide' ('Abide with me'); René François Armand Sully Prudhomme (1839–1907), the French poet and the recipient of the first Nobel prize in literature, in 1901; the Belgian poet Émile Cammaerts (1878–1953); and the Australian-born actor Leo McKern (1920–   ).

In 1976 the English actor Albert Finney played Hamlet in the first public performance at the National Theatre—almost twenty-five years after the foundation stone for the theatre had been laid.

✍ 17 *March*

THIS day is the festival of St Patrick (*c.*389–461), the patron saint of Ireland; and in 1497 a cave on an island in County Donegal, known as St Patrick's Purgatory from a revelation to the saint that it was an entrance to purgatory, was sealed by order of Pope Alexander VI. (It had for many years been visited by pilgrims in penitence, who believed they might witness both the torments of Hell and the joys of Heaven.)

Born on (and named from) St Patrick's Day were: Patrick Brontë (1777–1861), the curate of Haworth, in Yorkshire, and the father of the Brontë sisters; Sir Patrick Hastings (1880–1952), the British Labour MP and Attorney-General in the first Labour government, in 1924; and Patrick Hamilton (1904–61), the English playwright and novelist.

In 1912 Captain Lawrence Edward Grace Oates, on his 32nd birthday, deliberately walked out into a blizzard, knowing that his physical condition was impeding the progress of his companions on Scott's ill-fated return journey from the South Pole. ('I am just going outside, and may be some time.')

Also born on this day were: Ebenezer Elliott (1781–1849), the English poet known as 'the Corn-Law Rhymer'; the German inventor and motor car manufacturer Gottlieb Daimler (1834–1900); Margaret Bondfield (1873–1953), the first woman cabinet minister in Britain, in the second Labour government in 1929; Bobby Jones (1902–71), the American golfer who was the first to win the US and British Open Championships in the same year (1930); Robin Knox-Johnston (1939–   ), the British yachtsman who was the first to sail single-handed and non-stop around the world (in 1968–9); and the Russian-born ballet dancer and choreographer Rudolf Nureyev (1939–93).

❧ 18 March

Born on this day were: Sir William Cremer (1838–1908), the trade unionist, pacifist, and MP, who was the first Englishman to receive the Nobel peace prize (in 1903); Neville Chamberlain (1869–1940), the British prime minister who led his country into World War II ('In War, whichever side may call itself the victor, there are no winners, but all are losers.'); and Wilfred Owen (1893–1918), the English World War I soldier poet who died in action just one week before the Armistice. ('My subject is War, and the pity of War. The Poetry is in the pity.')

Also born on this day were: Stephen Grover Cleveland (1837–1908), the twenty-second and twenty-fourth US president and the only president to be elected for two separate terms of office; the French 'Symbolist' poet Stéphane Mallarmé (1842–98), whose poem 'L'Après-midi d'un faune' was illustrated by Manet and set to music by Debussy; the Russian composer Nicholas Andreievich Rimsky-Korsakov (1844–1908); Rudolf Diesel (1858–1913), the German engineer who designed and gave his name to the compression-ignition engine; and Frederik Willem de Klerk (1936–   ), the president of the Republic of South Africa who initiated the abolition of apartheid.

✌ 19 *March*

ON this day . . . in 1834 six agricultural labourers from the English village of Tolpuddle, in the county of Dorset, were sentenced to seven years' transportation for 'administering unlawful oaths'. (They were pardoned two years later, following nationwide protests, and as 'the Tolpuddle Martyrs' have become a symbol of the resistance of trade unionists against powerful employers.)

Born on this day were: David Livingstone (1813–73), the Scottish missionary and explorer; Sir Richard Burton (1821–90), the English explorer and orientalist who was the first Englishman to visit Mecca and the first to translate *The Arabian Nights* into English; Alfred von Tirpitz (1849–1930), the Prussian admiral who commanded the German fleet for the first part of World War I and whose name was given to a battleship of World War II; Sergei Pavlovich Diaghilev (1872–1929), the Russian ballet master and impresario; and the English composer Dame Elizabeth Maconchy (1907–   ).

✌ 20 *March*

TWO famous singers were born on this day and in the same year: the Italian tenor Beniamino Gigli (1890–1957); and the Danish tenor Lauritz Melchior (1890–1973).

Also born on this day were: Henrik Ibsen (1828–1906), the Norwegian dramatist and poet; the English actor Sir Michael Redgrave (1908–85); the Russian pianist Sviatoslav Richter (1915–   ); and Dame Vera Lynn (1917–   ), the English singer known in World War II as 'the Forces' Sweetheart'.

In 1549 Thomas Seymour, Baron Seymour of Sudeley, the fourth husband of Henry VIII's widow Catherine Parr and brother of Henry's third wife Jane Seymour, was executed on Tower Hill for treason. (He had had intentions of marrying the Princess Elizabeth.)

✅ 21 *March*

ON this day ... in 1556 Thomas Cranmer, the archbishop of Canterbury, was burnt at the stake in Oxford, having retracted the last of seven recantations that same day. (It is recorded that he first put his right hand to the flames, the hand with which he had signed his recantations, crying: 'This hath offended! Oh, this unworthy hand!')

Born on this day were: Johann Sebastian Bach (1685–1750), the German composer and patriarch of one of the great families of musicians; the Russian composer Modest Petrovich Mussorgsky (1839–81); Albert Chevalier (1861–1923), the British music-hall artist best known for his own song 'My Old Dutch'; Raoul Lufbery (1885–1918), the French-born American fighter pilot of World War I who was killed in combat; the French cellist and composer Paul Tortelier (1914–90); and the English composer and writer on music Antony Hopkins (1921–   ).

✅ 22 *March*

ON this day ... in 1919 the first international airline service was inaugurated on a weekly schedule between Paris and Brussels.

Born on this day were: Sir Anthony Van Dyck (1599–1641), the Flemish artist who as court painter to Charles I gave his name to the Vandyke beard, the short pointed beard depicted in his portraits of the English king; the German violinist and impresario Karl August Nicolas Rose (1842–89), who as 'Carl Rosa' founded (in England) the famous opera company named after him; the English novelist Nicholas Monsarrat (1910–79), whose most successful novel *The Cruel Sea* was inspired by his war service in the Royal Navy; Gerard Hoffnung (1925–59), the English artist, author, and broadcaster who founded the satirical 'Hoffnung Musical Festival'; the American song composer and lyricist Stephen Sondheim (1930–   ); and the English song composer Sir Andrew Lloyd Webber (1948–   ).

℘ 23 March

On this day ... in 1743, at the first performance in London of Handel's oratorio *Messiah*, George II and then everyone else in the audience at the Covent Garden Theatre rose to their feet at the beginning of the 'Hallelujah Chorus', thus establishing the traditional ceremony.

Born on this day were: William Smith, 'the father of English geology' (1769–1839), who published the first geological map of England and Wales; the Scottish painter Sir Muirhead Bone (1876–1953), who was the first official war artist (in 1916); Wernher von Braun (1912–77), the German-born American rocket pioneer who led the development of the V2 rockets used against Britain in 1944 and later initiated American space exploration; Donald Campbell (1921–67), the British racing motorist who followed his father Sir Malcolm Campbell in establishing land and water-speed records; and Sir Roger Bannister (1929–   ), the first man to run the mile in less than four minutes (in 1954, at Oxford).

℘ 24 March

On this day ... in 1603 Elizabeth I died ('All my possessions for a moment of time') and James VI of Scotland, whose mother Mary, Queen of Scots, had been executed with Elizabeth's consent in 1587, became king of Great Britain as James I.

In 1916 the Spanish composer Enrique Granados and his wife died when the ship on which they were returning from America (where they had attended the first performance of his opera *Goyescas*) was torpedoed by a German submarine in the English Channel.

Born on this day were: the American hymn-writer Fanny Crosby (1820–1915), who in spite of being blind from infancy wrote more than 6,000 compositions; William Morris (1834–96), the English poet, artist, and socialist whose name has become synonymous with a type of decorative art; and the English author and broadcaster Malcolm Muggeridge (1903–90).

ᘐ 25 March

ON this day . . . in 1957 the EEC (the European Economic Community), known popularly as the Common Market, was established by the Treaty of Rome, which was signed by representatives of Belgium, France, Federal Germany, Italy, Luxembourg, and The Netherlands ('Les Six') and ratified the following year.

Born on this day were: Joachim Murat (1767–1815), Napoleon's brother-in-law who became king of Naples (in 1808), but was executed after Napoleon's defeat at Waterloo; the Italian conductor Arturo Toscanini (1867–1957); the Hungarian composer Béla Bartók (1881–1945); the English composer Haydn Wood (1882–1959), best remembered for his ballads (especially 'Roses of Picardy'); A(lan) J(ohn) P(ercivale) Taylor (1906–90), the English historian who pioneered the presentation of the history lecture on British television; and Sir David Lean (1908–91), the British film director.

In 1975 King Faisal of Saudi Arabia was assassinated by a deranged kinsman.

ᘐ 26 March

ON this day . . . in 1827 the German composer Ludwig van Beethoven died in Vienna, having been almost completely deaf for the latter part of his life. ('I shall hear in heaven.')

Born on this day were: A(lfred) E(dward) Housman (1859–1936) the English poet especially remembered for *A Shropshire Lad*; the American poet Robert Frost (1874–1963), who was awarded the Pulitzer prize four times (in 1923, 1930, 1936, and 1942); Tennessee Williams (1911–83), the American playwright who was awarded the Pulitzer prize for his plays *A Streetcar Named Desire* (in 1947) and *Cat on a Hot Tin Roof* (in 1955); the English novelist Elizabeth Jane Howard (1923–); Pierre Boulez (1925–), the French composer and conductor; and the Korean violinist Kyung-Wha Chung (1948–).

ᏇᎦ 27 *March*

ON this day . . . in 1899 the first international radio transmission between England and France was achieved by the Italian inventor Guglielmo Marconi. (The message, transmitted in Morse code, was also the first press communication to be sent from one country to another by radio.)

Born on this day were: Wilhelm Konrad von Röntgen (1845–1923), the German scientist who discovered X-rays (originally named after him as Röntgen rays); Sir Henry Royce (1863–1933), the founder (with the Hon. C. S. Rolls) of Rolls-Royce Ltd. in 1904; the German novelist Heinrich Mann (1871–1950), the brother of the novelist Thomas Mann; the American composer Ferde Grofé (1892–1972), best remembered for his orchestration of Gershwin's *Rhapsody in Blue*; Lord Callaghan (1912–   ), the British prime minister as James Callaghan 1976–9; and the Russian cellist and conductor Mstislav Rostropovich (1927–   ).

ᏇᎦ 28 *March*

SIR William Howard Russell (1820–1907), the British war correspondent who coined the phrase 'the thin red line' ('the Russians dash on towards that thin red line tipped with steel'), was born on this day, the day that the Crimean War began in 1854.

In World War I, in 1917, the Women's Army Auxiliary Corps (the WAAC) was founded. (These were Britain's first official servicewomen.)

In World War II, in 1941, the Italian fleet was routed by the British at the battle of Matapan; in 1942 a commando raid, 'Operation Chariot', was carried out on Saint-Nazaire, in France; and in 1945 the last of the V2 rockets landed in England.

Also born on this day were: George I (1660–1727), the first Hanoverian king of Great Britain; Aristide Briand (1862–1932), the French statesman and recipient of the 1926 Nobel peace prize; the American bandleader and impresario Paul Whiteman (1890–1967); the British actress Dame Flora Robson (1902–84); the British actor and writer Dirk Bogarde (1921–   ); and the British Labour politician Neil Kinnock (1942–   ).

In 1941 the English writer Virginia Woolf committed suicide by drowning.

᪥ 29 *March*

ON this day . . . in 1461 Edward IV secured his claim to the English throne in defeating Henry VI's Lancastrians at the battle of Towton, in the West Riding of Yorkshire ('The bloodiest battle on English soil'); and in 1644 Sir William Waller's parliamentarians defeated the royalists under Lord Hopton at the battle of Cheriton, in Hampshire.

Born on this day were: John Tyler (1790–1862), the tenth US president and the first vice-president to succeed to office on the death of the president (W. H. Harrison, in 1841); the Earl of Derby (1799–1869), the British prime minister named by another prime minister, Disraeli, as 'the Rupert of parliamentary discussion . . . in his charge he is relentless, but when he returns from the pursuit he always finds his camp in the possession of the enemy'; Elihu Thomson (1835–1937), the English-born American inventor of electric welding and arc lighting; Sir Edwin Landseer Lutyens (1869–1944), the architect of the Cenotaph and Liverpool's Roman Catholic cathedral; the English composer Sir William Walton (1902–83); the English composer Richard Rodney Bennett (1936–   ); and John Major (1943–   ), the British prime minister since 1990.

In 1912 Captain Robert Falcon Scott wrote the last entry in his diary: 'It seems a pity, but I do not think I can write any more.'

᪥ 30 *March*

TWO of the world's greatest artists were born on this day: the Spanish painter Francisco José de Goya y Lucientes (1746–1828); and the Dutch painter Vincent Willem Van Gogh (1853–90).

Also born on this day were: Anna Sewell (1820–78), the English novelist whose *Black Beauty* has become the classic story about horses; Paul Verlaine (1844–96), the French poet and originator of the 'Decadent' school of literature; the Irish playwright Sean O'Casey (1884–1964); and the English poet Julian Grenfell (1888–1915), who was killed in action in World War I.

'Beau' Brummell, the English dandy and former favourite of the Prince Regent, died in a French lunatic asylum for paupers in 1840.

₰ 31 March

ON this day . . . in 1282 the great massacre of the French in Sicily ('the Sicilian Vespers') came to an end. (The tolling of the bell for vespers the day before had been the signal for the start of the massacre.)

Born on this day were: René Descartes (1596–1650), the French philosopher and scientist known as 'the father of modern philosophy'; the English poet Andrew Marvell (1621–78), known as 'the incorruptible patriot'; John Harrison (1693–1776), the Englishman who invented the chronometer; the Austrian composer Joseph Haydn (1732–1809); the Russian writer Nikolai Vasilievich Gogol (1809–52); the English writer Edward Fitzgerald (1809–83), who translated the quatrains of the eleventh-century Persian poet and astronomer Omar Khayyám (*The Rubáiyát of Omar Khayyám*); the German scientist Robert Wilhelm Bunsen (1811–99), who invented and gave his name to the bunsen burner; Arthur Griffith (1872–1922), the first president of the Irish Free State; Jack Johnson (1878–1946), the first black American boxer to become the world heavyweight champion; and Sir William Lawrence Bragg (1890–1971), the first Nobel prizewinner to share the prize with his father (in 1915, for physics); and Al Gore (1948–   ), the US vice-president.

APRIL ✤

✤ 1 April

On this day . . . in 1939 the Spanish Civil War was effectively ended with the official recognition of Franco's government by the USA.

Born on this day were: the Italian composer and pianist Ferruccio Busoni (1866–1924); the Russian composer and pianist Sergei Rachmaninov (1873–1943); and the English composer and pianist Stephen Russell 'Steve' Race (1921–   ).

Also born on this day were: William Harvey (1578–1657), the English physician who discovered the circulation of the blood; Antoine François Prévost d'Exiles, known as Abbé Prévost (1697–1763), the French writer whose novel *Manon Lescaut* was adapted as operas by Auber, Massenet, and Puccini; William Mulready (1786–1863), the Irish painter who designed the first penny postage envelope (in 1840); the Prussian prime minister Prince Otto Eduard Leopold von Bismarck (1815–98), who founded the German Empire and was known as 'the Iron Chancellor' (from his policy involving military power: 'this policy cannot succeed through speeches . . . and songs; it can only be carried out through *blood and iron*'); and Edmond Rostand (1868–1918), the French poet and playwright, creator of *Cyrano de Bergerac*.

In 1918 the RNAS (the Royal Naval Air Service) and the RFC (the Royal Flying Corps) were amalgamated as the RAF (the Royal Air Force).

✤ 2 April

On this day . . . in 1801, at the battle of Copenhagen, the Danish fleet was defeated by the British. (Although officially second in command to Sir Hyde Parker, Nelson conducted the attack in his own way: putting his telescope to his right (blind) eye, he said, 'I have a right to be blind sometimes . . . I really do not see the signal!') In 1982 Argentina invaded the Falkland Islands, beginning the conflict which ended on 14 June of the same year with the surrender of the Argentinian forces.

Born on this day were: Giovanni Jacopo Casanova (1725–98), the Italian writer and adventurer whose name has become synonymous with that of the legendary Don Juan; August Heinrich Hoffmann von Fallersleben (1798–1874), the German poet whose 'Deutschland über Alles' became the German national anthem in 1841; Hans Christian Andersen (1805–75), the Danish poet and storyteller of national and international fame; William Holman Hunt (1827–1910), the English painter and Pre-Raphaelite; the French writer Émile Zola (1840–1902); the German painter Max Ernst (1891–1976), the founder of the 'Dadaist' movement in art; the British actor Sir Alec Guinness (1914–   ); and Sir Jack Brabham (1926–   ), the Australian-born world champion racing driver in 1959–60, 1960–1, and 1966.

✌ 3 April

THE English King Henry IV (1367–1413) was born on the same day that his father John of Gaunt and his uncle Edward, the Black Prince, were victorious at the battle of Najara, in Spain.

Also born on this day were: George Herbert (1593–1633), the English clergyman and poet and brother of Lord Herbert of Cherbury; Washington Irving, alias 'Diedrich Knickerbocker' (1783–1859), the American writer who created 'Rip Van Winkle'; Reginald De Koven (1859–1920), the American composer who wrote an opera entitled *Rip Van Winkle*; the Spanish composer Mario Castelnuovo-Tedesco (1895–1968), who wrote settings of all the songs from Shakespeare's plays; Marlon Brando (1924–   ), the American actor and one of the pioneers of 'the Method'; and the American actress and singer Doris Day (1924–   ).

In 1882 the American outlaw Jesse James was shot in the back and killed by one of his own gang, Robert Ford, for a $5,000 reward.

✌ 4 April

ON this day . . . in 1968 the American civil rights leader Martin Luther King was assassinated in Memphis, Tennessee; and in 1979 the president of Pakistan, Zulfikar Ali Bhutto, was executed.

Born on this day were: Grinling Gibbons (1648–1721), the Dutch-born English woodcarver and sculptor who was employed by Wren and others to execute his works in St Paul's Cathedral and other famous buildings; Linus Yale (1821–68), the American inventor of the cylindrical lock named after him; Sir William Siemens (1823–83), the German-born inventor and electrical engineer; Hans Richter (1843–1916), the Austro-Hungarian conductor and the dedicatee of Elgar's First Symphony; the French conductor Pierre Monteux (1875–1964), who at the age of 86 became the chief conductor of the London Symphony Orchestra with a twenty-five-year contract; the French 'Fauvist' painter Maurice de Vlaminck (1876–1958); and the Scottish painter Sir William Russell Flint (1880–1969).

৯ 5 April

On this day . . . in 1794 the French Revolutionary leader Georges Jacques Danton ended his political career on the guillotine. ('Be sure you show the mob my head. It will be a long time ere they see its like.') In 1955 the 80-year-old Sir Winston Churchill ended his effective political career rather less dramatically when he resigned the premiership. (Because of a strike, there were no British national newspapers to report the event.)

In 1923 Lord Carnarvon, who had discovered Tutankhamun's tomb a few months previously, died from an unidentified illness (or, as it was said, 'the curse of Tutankhamun').

Born on this day were: the English philosopher Thomas Hobbes (1588–1679), as the Spanish Armada was seen to be approaching England; Louis Spohr (1784–1859), the German composer believed to have been the first to conduct with a baton; Joseph Lister, later Lord Lister (1827–1912), the British surgeon who was the first to use the antiseptic method of treating wounds; the English poet Algernon Charles Swinburne (1837–1909); and the Austrian conductor Herbert von Karajan (1908–89).

✱ 6 April

ON this day . . . in 1199 the English king Richard I, known as 'Cœur de Lion', was killed by an arrow at the siege of the castle of Chaluz, near Limoges in France. (It is believed that in all his life he had not spent a full year in England; and according to the authors of *1066 and All That* was 'therefore known as Richard Gare de Lyon'.)

In 1896 the first modern Olympic Games began in Athens; in 1909 the American explorer Robert Edwin Peary became the first man to reach the North Pole; and in 1917 the USA declared war on Germany.

Raphael (Raffaello Sanzio), the Italian painter, died in 1520. (It is uncertain whether this day was not also his day of birth, in 1483.)

Born on this day were: James Mill (1773–1836), the Scottish philosopher and father of the philosopher John Stuart Mill; the American escapologist Harry Houdini (1874–1936), who took his stage-name from the French illusionist 'Houdin'; and the Dutch aircraft designer Anthony Fokker (1890–1939), who invented the first effective device for firing a machine-gun through a rotating propeller.

✱ 7 April

ON this day . . . in 1862 the second major engagement of the American Civil War, the battle of Shiloh, came to an end (it had begun the previous day) with a victory for the Union army. In 1945 the Japanese battleship *Yamato* was sunk off Japan by US aircraft. (It was the world's largest battleship and had never been in action against another warship.)

Born on this day were: the Spanish missionary St Francis Xavier (1506–52); Dr Charles Burney (1726–1814), the English musicologist and the father of the novelist Fanny Burney; William Wordsworth (1770–1850), the English Poet Laureate from 1843; Charles Fourier (1772–1837), the French socialist who gave his name to the social theory Fourierism; Sir David Low (1891–1963), the New Zealand-born cartoonist who created 'Colonel Blimp'; Billie Holiday (1915–59), the American jazz singer known as 'Lady Day'; and the Indian sitar player and composer Ravi Shankar (1920–   ).

In 1739 the notorious English highwayman Dick Turpin was hanged on the Knavesmire, at York. (His identity had been revealed through his handwriting when he was in prison for a minor offence.)

✌ 8 April

ON this day... in 1838, the first regular transatlantic steamship service for passengers began as the *Great Western*, designed by Isambard Kingdom Brunel, set out on her maiden voyage from Bristol to New York. (The journey took some fifteen days, which was half the time taken by the fastest sailing ship.)

Born on this day were: Giuseppe Tartini (1692–1770), the Italian violinist and composer said to be the greatest instrumentalist of his day; Jem Mace (1831–1910), the British pugilist known as 'the Gypsy' whose career in boxing lasted some thirty-five years; Sir Adrian Boult (1889–1983), the English conductor who was still conducting a year or so before his death at the age of 93; and Mary Pickford (1893–1979), the American film actress who became one of the world's richest women.

In 1898 General Kitchener's forces defeated the Khalifa, the leader of the dervishes in Sudan, at the battle of Atbara.

✌ 9 April

ON this day ... in 1917, in World War I, Canadian troops began their massive assault on Vimy Ridge, near Arras in France (capturing it the next day); and the English poet Edward Thomas was killed in action at Arras on the same day.

Born on this day were: James, Duke of Monmouth (1649–85), a natural son of Charles II who led and gave his name to the unsuccessful rebellion against James II in 1685; Isambard Kingdom Brunel (1806–59), the English railway and marine engineer who built the Clifton suspension bridge; Charles Baudelaire (1821–67), the French poet, and one of the 'Decadents'; Sir Paolo Tosti (1846–1916), the Italian composer who became a naturalized British subject and singing teacher to the British royal family; Adela Florence Nicolson (1865–1904), the English poet who used the pseudonym of 'Laurence Hope' for her best-known *Indian Love Lyrics*; Erich von Ludendorff (1865–1937), the German general and chief-of-staff to Hindenburg in World War I; and Paul Robeson (1898–1976), the American singer and actor.

In 1747 the Scottish Jacobite Lord Lovat was beheaded on Tower Hill for high treason. (He was the last man to be executed by decapitation in Britain.)

✌ 10 *April*

ON this day . . . in 1814 the defeat of Marshal Soult by Wellington at the battle of Toulouse led to the abdication of Napoleon the following day; and in 1912 the ill-fated White Star liner *Titanic* embarked on her maiden voyage.

Born on this day were: the English essayist and critic William Hazlitt (1778–1830); Lew Wallace (1827–1905), the American writer best known for his novel *Ben Hur*, who was also a distinguished general in the Union army; 'General' William Booth (1829–1912), the English founder of the Salvation Army; Joseph Pulitzer (1847–1911), the Hungarian-born newspaper proprietor who founded and gave his name to prizes for literature, drama, music, and journalism which were confined to American citizens; and Ben Nicholson (1894–1982), the English sculptor who was married to fellow sculptor Dame Barbara Hepworth.

✌ 11 *April*

ON this day . . . in 1713 the Treaty of Utrecht ended the War of the Spanish Succession, the alliance of Britain, Holland, Austria, Prussia, and the Holy Roman Empire against Louis XIV of France; and in 1814 Louis XVIII, the younger brother of Louis XVI, became king of France on the abdication of Napoleon.

Born on this day were: George Canning (1770–1827), the British prime minister who died after only a few months in office; Sir Charles Hallé (1819–95), the German-born pianist and conductor who founded the Hallé Orchestra in Manchester; Nick La Rocca (1889–1961), the

American cornettist who founded and led the Original Dixieland Jazz Band, the first jazz band to make gramophone records; and the English painter John Nash (1893–1977), the younger brother of the painter Paul Nash.

In 1945 the German concentration camp at Buchenwald was the first of the notorious camps to be liberated by the Allies in the last month of World War II.

৯৯ 12 *April*

ON this day ... in 1782 Admiral Rodney commanded the British fleet in the battle of the Saints (off Dominica), the only major naval battle won by Britain in the American War of Independence; and in 1861 the American Civil War began with a Confederate attack on Fort Sumter, in Charleston Harbor.

In 1911 a Frenchman, Pierre Prier, made the first non-stop flight from London to Paris, his journey taking almost four hours; and in 1961 Russian cosmonaut Yuri Gagarin became the first man to orbit the earth, his journey lasting 108 minutes.

Henry Clay (1777–1852), the American secretary of state 1825–9, who was three times an unsuccessful candidate for the presidency, was born on this day; and Franklin Delano Roosevelt, who was elected as the US president for a record four times, died in office in 1945, in the year after his last election.

৯৯ 13 *April*

BORN on this day were: Catherine de Medici (1519–89), the consort of Henri II of France and the mother of three other kings of France; Lord North (1732–92), the British prime minister during the American War of Independence; Thomas Jefferson (1743–1826), the American president who drafted the Declaration of Independence; Joseph Bramah (1748–1814), the English inventor who gave his name to a safety lock and a hydraulic press; Richard Trevithick (1771–1833), the English inventor who built and demonstrated the first steam-propelled passenger vehicle:

the English composer William Sterndale Bennett (1816–75), who founded the Bach society and conducted the first performance in England of the *St Matthew Passion* (in 1854); the Scottish composer Sir John McEwen (1868–1948), whose 'Solway' Symphony was the first symphony by a British composer to be recorded; Sir Arthur 'Bomber' Harris (1892–1984), the commander-in-chief of RAF Bomber Command in World War II, 1942–5; Sir Robert Watson-Watt (1892–1973), the Scottish scientist who was mainly responsible for the development of radar in World War II; and the Irish playwright Samuel Beckett (1906–89), the recipient of the 1969 Nobel prize for literature.

🈂 14 *April*

On this day . . . in 1471 the Earl of Warwick, 'the Kingmaker', who had fought for both sides since the beginning of the Wars of the Roses, was killed at the battle of Barnet with the defeat of the Lancastrians and the subsequent restoration of Edward IV to the throne of England.

In 1865 the US president Abraham Lincoln was shot by a deranged actor, John Wilkes Booth, at Ford's Theater in Washington, DC, and died the following morning; and in 1912 the Atlantic passenger liner *Titanic*, on her maiden voyage and hailed as 'the unsinkable ship', struck an iceberg and sank on the following day with the loss of more than 1,500 lives.

Born on this day were: Christian Huygens (1629–95), the Dutch scientist who invented the pendulum; Arnold Joseph Toynbee (1889–1975), the English historian who was the nephew of Arnold Toynbee the historian and social reformer; and the English actor Sir John Gielgud (1904–   ), who is a grand-nephew of the actress Dame Ellen Terry.

🈂 15 *April*

Born on this day were: Leonardo da Vinci (1452–1519), the Italian artist and scientist who, in making an application to the Duke of Milan for the position of city administrator, gave his curriculum vitae as 'painter, architect, philosopher, poet, composer, sculptor, athlete, mathematician, inventor, and anatomist'; Sir James Clark Ross (1800–62), the Scottish explorer who located the North Magnetic Pole in 1831; Théodore Rousseau (1812–67), the French landscape painter known as 'le grand refusé' (from his lack of success in getting his paintings accepted for exhibition); the German artist Wilhelm Busch (1832–1908), the originator of the strip cartoon; the American novelist Henry James (1843–1916), who became a naturalized British citizen in 1915; Bessie Smith (1894–1937), the American jazz singer known as 'the Empress of the Blues'; and Jeffrey Archer (Lord Archer), the British author and former politician (1940–   ).

In 1912 the *Titanic* sank (see 14 April) with its lights blazing and the ship's band playing. (Some accounts say that their last rendition was the hymn 'Nearer, My God to Thee', while others maintain that it was a rag-time tune of the day.)

⥁ 16 *April*

On this day... in 1746 the Young Pretender ('Bonnie Prince Charlie') was defeated at the battle of Culloden, the final major attempt at returning the Stuarts to the throne of England and the last pitched battle fought in Britain.

Born on this day were: Sir Hans Sloane (1660–1753), the British physician and naturalist whose library and natural history collection became the foundation of the British Museum; Sir John Franklin (1786–1847), the Arctic explorer who discovered the North-West Passage but perished on his final expedition; the Scottish bookseller and printer William Chambers (1800–83), who (with his brother Robert) founded the publishing firm of W. & R. Chambers; Ford Madox Brown (1821–93), the British historical painter; the American airman Wilbur Wright (1867–1912), who, with his brother Orville, was the first man to fly in a heavier-than-air machine; Sir Charles Spencer 'Charlie' Chaplin (1889–1977), the English-born film actor and director; Spike Milligan (1918–   ), the British writer and comedian, one of the Goons of the radio series of the 1950s; Sir Peter Ustinov (1921–   ), the British actor, dramatist, and novelist; and the English poet and novelist Sir Kingsley Amis (1922–   ).

✒ 17 *April*

ON this day . . . in 1521 Martin Luther appeared before the Diet of Worms prior to his excommunication.

Born on this day were: Henry Vaughan (1622–95), the Welsh religious poet known as 'the Silurist' (from the ancient tribe of Silures who inhabited South Wales); Constantine P. Kavatis, the Greek poet known as 'Cavafy' (1863–1933); the Anglo-Irish writer known as Robert Tressell (1870–1911); the Scottish novelist and playwright Ian Hay (1870–1952), who coined the phrase 'funny peculiar, or funny ha-ha?' and whose original name was John Hay Beith; the Austrian pianist Artur Schnabel (1882–1951); Nikita Khrushchev (1894–1971), the Soviet premier 1958–64; Thornton Wilder (1897–1975), the American novelist and playwright whose novel *The Bridge of San Luis Rey* gave him the first of his three Pulitzer prizes; and Mrs Sirimavo Bandaranaike (1916–   ), the prime minister of Ceylon (from 1972 Sri Lanka) who was, in 1960, the first woman prime minister.

✒ 18 *April*

ON this day . . . in 1775 an American silversmith who had taken part in the 'Boston Tea Party' in 1773 made his famous ride from Boston to Lexington to warn of the coming of the British Army, immortalized in Longfellow's poem 'Paul Revere's Ride'. ('Listen, my children, and you shall hear | Of the midnight ride of Paul Revere, | On the eighteenth of April in seventy-five . . .')

Born on this day were: Lucrezia Borgia (1480–1519), a daughter of Pope Alexander VI and a patroness of the arts, who has been accused (perhaps wrongly) of poisoning her adversaries; George Henry Lewes (1817–78), the English philosopher and writer and lover of George Eliot; the Austrian composer Franz von Suppé (1819–95); Clarence Darrow (1857–1938), the American lawyer famed for his defence of the murderers Leopold and Loeb in 1924, and in the following year the schoolteacher John Scopes in the so-called 'monkey trial' concerning the teaching of the Darwinian theory of evolution in a public school; the aviation pioneer Humphrey Verdon Roe (1878–1949), who founded (with his brother (Sir) Alliot Verdon-Roe) the AVRO aircraft company and (with his wife Dr Marie Stopes) the first birth control clinic in Britain; and Leopold Stokowski (1882–1977), the British-born American conductor.

In 1689 Lord Jeffreys, known as 'the hanging judge', who had sentenced hundreds of men to death following the Monmouth Rebellion in 1685, died of dissipation in the Tower of London. (He had been captured attempting to flee the country disguised in woman's clothes, after the abdication of his patron James II.)

✌ 19 April

ON this day . . . in 1775 the first battle of the American War of Independence began at Lexington, Massachusetts, with the British general Thomas Gage sending a force to seize military weapons at Concord; and in 1880 the war correspondent of *The Times* telephoned his report of the battle of Ahmed Khel (fought on this day during the Second Afghan War), which was published in London via telegraph and other communications on the following morning. (This was the first time that news had been sent from a field of battle in this way.)

Born on this day were: the French composer Germaine Tailleferre (1892–1983), who was one of a group of French composers known as 'Les Six'; Herbert Wilcox (1892–1977), the British film producer and director who was married to the actress Dame Anna Neagle; and the British actor, comedian, and musician Dudley Moore (1935–   ).

In 1824 Lord Byron died at Missolonghi, in Greece. (He was on his way to assist the Greeks in their struggle against the Turks when he was struck down by malaria.) The former British prime minister Benjamin Disraeli, Lord Beaconsfield, died in 1881; and the anniversary has since been known as 'Primrose Day', from the wreath of primroses sent to his funeral by Queen Victoria.

✌ 20 April

ON this day . . . in 1657 the English admiral Robert Blake fought his last battle when he destroyed the Spanish fleet in Santa Cruz Bay. (He died as his ship entered Plymouth Harbour on 7 August of that year.) In 1770 another English seafarer, Captain James Cook, discovered New South Wales.

Born on this day were: Charles Louis Napoleon Bonaparte, Napoleon III (1808–73), Napoleon's nephew and emperor of France 1852–70; Adolf Hitler (1889–1945), the German dictator; the Spanish painter Joan Miró (1893–1983); and the American film comedian Harold Lloyd (1894–1971).

21 April

ON this day . . . in 1509 the English king Henry VII died and was succeeded by his second son Henry VIII; and in 1918 Baron von Richthofen, the German fighter pilot ace of World War I (known as 'the Red Baron', from the colour of his Fokker triplane), was shot down and killed, and succeeded as squadron commander by one Hermann Goering.

Born on this day were: Friedrich Froebel (1782–1852), the German educationalist who founded and gave his name to the kindergarten system of education; Charlotte Brontë (1816–55), the eldest of the Brontë sisters; Dr Richard Beeching, later Lord Beeching (1913–85), the chairman of British Railways, 1963–5, whose name has come to be used as a synonym for 'axing' unprofitable services, the 'Beeching axe'; and Her Majesty Queen Elizabeth II (1926–   ), who, like Elizabeth I, was in her 26th year when she became queen.

22 April

THE Russian revolutionary leader Nikolai Lenin (whose original name was Vladimir Ilich Ulyanov) was born on this day in 1870. (Petrograd was renamed Leningrad on this day in 1924, the year of Lenin's death, but reverted to its original name of St Petersburg in 1991.)

Also born on this day were: Henry Fielding (1707–54), the English novelist and creator of *Tom Jones*; the German philosopher Immanuel Kant (1724–1804); Mme de Staël (1766–1817), one of the leading French women writers of her time; Phil May (1864–1903), the English illustrator and caricaturist and *Punch* cartoonist; J. Robert Oppenheimer (1904–67), the American scientist known as 'the father of the atomic bomb' (the 'J.' in his name stands for no particular name); Eric Fenby (1906–   ), the English composer who was the amanuensis to Frederick Delius; Kathleen Ferrier (1912–53), the English contralto whose tragically short career began in her 30th year; Sir Yehudi Menuhin (1916–   ), the American-born violinist who made his professional début at the age of 7; and the British architect Sir James Stirling (1926–92).

In 1915 the second battle of Ypres began and the Germans first used poison gas against British troops; and in 1918 the battle of Zeebrugge began, as British forces commanded by Admiral Keyes attempted to sink block-ships in the German U-boat bases.

❧ 23 April

Two of the world's greatest writers died on this day in 1616: the Spanish novelist, playwright, and poet Miguel de Cervantes Saavedra (the creator of *Don Quixote*); and the English playwright and poet William Shakespeare, who it is believed was also born on this day in 1564.

Other famous Englishmen (and one Englishwoman) born on this day (St George's Day, after the patron saint of England, 'adopted' by Edward III) were: George Anson, Lord Anson (1697–1762), circumnavigator of the world 1740–4; Samuel Wallis (1728–95), another circumnavigator and the discoverer of Tahiti, in 1767; the landscape painter J(oseph) M(allord) W(illiam) Turner (1775–1851); Dame Ethel Smythe (1858–1944), the first composer to receive the DBE (in 1922); and Lord Allenby (1861–1936), the commander-in-chief of the Egyptian Expeditionary Force which captured Jerusalem from the Turks in 1917.

Also born on this day were: James Buchanan (1791–1868), the fifteenth US president and the only bachelor to hold that office; Max Planck (1858–1947), the German scientist who originated the quantum theory; the New Zealand-born novelist Dame Ngaio Marsh (1899–1982); the Russian-born novelist Vladimir Nabokov (1899–1977); the Irish-American novelist and playwright James Patrick Donleavy (1926–   ); and Shirley Temple (1928–   ), the former child star of Hollywood who became an American diplomat as Shirley Temple Black.

In 1915 the English poet Rupert Brooke, famed for his war poem 'If I should die, think only this of me . . .', died of a fever on the Greek island of Skyros, *en route* to the Dardanelles.

‹ð 24 *April*

ON this day . . . in 1792 Claude Joseph Rouget de Lisle, a young French engineer officer stationed at Strasbourg, began composing the words and music of what was to become the French national anthem, 'La Marseillaise'. (It was originally entitled 'Chant de guerre pour l'Armée du Rhin', but took its eventual title after being adopted by soldiers at Marseilles who were about to march on Paris.) Born on this day was Arthur Christopher Benson (1862–1925), a son of the archbishop of Canterbury Edward White Benson and author of the poem 'Land of Hope and Glory', which, set to music by Sir Edward Elgar, became regarded as the 'second' British national anthem.

Also born on this day were: Edmund Cartwright (1743–1823), the English parson who invented the power loom; the English novelist Anthony Trollope (1815–82); Henri Philippe Pétain (1856–1951), the French marshal known in World War I as 'the hero of Verdun' who was subsequently convicted of treason in 1945; Air Chief Marshal Hugh Dowding, Lord Dowding (1882–1970), the commander-in-chief of RAF Fighter Command during the battle of Britain, who was known as 'Stuffy Dowding'; and William Joyce (1906–46), the British traitor known as 'Lord Haw-Haw' from his anti-British broadcasts from Berlin in World War II.

In 1916 a rebellion organized by Sinn Fein (the Irish nationalist movement), 'the Easter Rebellion', began on Easter Monday in Dublin. ('All changed, changed utterly: | A terrible beauty is born.'—W. B. Yeats.)

‹ð 25 *April*

BORN on this day were: Edward II (1284–1327), king of England 1307–27, and the first heir apparent to be created Prince of Wales (in

1301); and Oliver Cromwell (1599–1658), the first Lord Protector of England, 1653–58.

Also born on this day were: Sir Marc Isambard Brunel (1769–1849), the French-born engineer and inventor; John Keble (1792–1866), the English clergyman and poet whose name is commemorated by Oxford's Keble College; C(harles) B(urgess) Fry (1872–1956), the Oxford University, Sussex, and England cricketer who also represented his country in soccer and athletics and held the world record for the long jump; the English poet Walter de la Mare (1873–1956); Guglielmo Marconi (1874–1937), the first Italian to be awarded the Nobel physics prize (in 1909); and the American jazz singer Ella Fitzgerald (1918–   ), known as 'the first lady of song'.

In 1915, in the second year of World War I, the Allied operations in Gallipoli began with the landing of the Australian and New Zealand Army Corps. (The anniversary has since been commemorated as Anzac Day.)

🕮 26 April

On this day . . . in 1865 John Wilkes Booth, the American actor who had assassinated Abraham Lincoln on 14 April, was shot dead (probably by his own hand) after being cornered in a blazing building.

Born on this day were: John James Audubon (1785–1851), the American ornithologist and painter who illustrated all known species of North American birds in his celebrated ten-volume *Birds of America*; the French painter Ferdinand Victor Eugène Delacroix (1798–1863); the British aviation pioneer Sir Alliot Verdon-Roe (1877–1958), who built the first successful British powered aeroplane and founded the company which produced the famous AVRO aircraft; Anita Loos (1893–1981), the American novelist and screenwriter (*Gentlemen Prefer Blondes*); Rudolf Hess (1894–1987), the German war criminal who flew alone to Britain in 1941 to negotiate an Anglo-German 'peace treaty'; and Charles Francis Richter (1900–85), the American seismologist who gave his name to the scale for measuring the intensity of earthquakes.

In 1915 Second-Lieutenant Rhodes-Moorhouse of the Special Reserve Flying Corps became the first airman to win the Victoria Cross after carrying out a successful bombing raid. (Although mortally wounded, he returned in his aircraft and gave a report on his mission.)

❧ 27 April

ON this day . . . in 1296 the Scots were defeated by Edward I at the battle of Dunbar.

In 1749 the first official performance of Handel's *Royal Fireworks Music* (celebrating the Treaty of Aix-la-Chapelle the previous year) was curtailed by the outbreak of a fire.

Born on this day were: Edward Gibbon (1737–94), the English historian who wrote *The Decline and Fall of the Roman Empire*; Mary Wollstonecraft Godwin (1759–97), who wrote *Vindication of the Rights of Women* and was the mother of Mary Shelley; Samuel Finley Breese Morse (1791–1872), the American artist and inventor of the Morse code; Herbert Spencer (1820–1903), the English philosopher whose *Principles of Psychology* is said to have anticipated Darwin's theory of evolution; Ulysses Simpson Grant (1822–85), the commander of the Union armies in the American Civil War and afterwards the eighteenth US president; Edward Whymper (1840–1911), the English wood-engraver and mountaineer who was the first to scale the Matterhorn; Cecil Day Lewis (1904–72), the English Poet Laureate 1968–72; and the British airwoman Sheila Scott (1927–88).

Ferdinand Magellan, the Portuguese navigator who discovered and gave his name to the strait separating the Atlantic and the Pacific, was killed on his last voyage in 1521. (This resulted in the first circumnavigation of the world sailing westward by Juan Sebastian del Cano, one of his commanders, and seventeen surviving crewmen from the original expedition of five ships.)

❧ 28 April

ON this day . . . in 1789 the notorious 'Mutiny on the *Bounty*' reached its climax as the crew of HMS *Bounty*, led by the master's mate Fletcher Christian, overpowered the ship's captain William Bligh and cast him adrift with eighteen others in the ship's boat; and in 1947 the Norwegian anthropologist Thor Heyerdahl and five others set out in a balsa-wood raft, *Kon-Tiki*, to an island in the South Pacific to prove that Peruvian Indians could have settled in Polynesia.

Born on this day were: Edward IV (1442–83), king of England 1461–83, and the first king of the House of York; James Monroe (1758–1831), the fifth US president and instigator of the Monroe Doctrine which implied that the 'New World' would avoid involvement with wars of the 'Old World'; Lord Shaftesbury (1801–85), the English philanthropist and reformer commemorated by the statue of *Eros* in London's Piccadilly Circus; the American actor Lionel Barrymore (1878–1954), brother of the actress Ethel Barrymore and the actor John Barrymore; James Baker (1930–   ), the US secretary of state 1988–92; and Saddam Hussein (1937–   ), the president of Iraq since 1979.

In 1945 the Italian dictator Benito Mussolini and his mistress Claretta Petucci were shot dead by Italian partisans.

The League of Nations was founded in 1919.

ஃ 29 *April*

THREE famous 'bandleaders' were born on this day: Sir Thomas Beecham (1879–1961), the English conductor and impresario who was the grandson of Thomas Beecham (the inventor of 'Beecham's Pills') and the founder of the London Philharmonic Orchestra (in 1932) and the Royal Philharmonic Orchestra (in 1946); the English conductor Sir Malcolm Sargent (1895–1967); and Edward Kennedy 'Duke' Ellington (1899–1974), the American musician described by the English composer and conductor Constant Lambert as 'no mere bandleader and arranger, but a composer of uncommon merit, probably the first composer of character to come out of America'.

Also born on this day were: Arthur Wellesley, the 1st Duke of Wellington (1769–1852), the British soldier and prime minister; Tsar Alexander II (1818–81); the American newspaper publisher William Randolph Hearst (1863–1951), whose life was the subject of Orson Welles's famous film *Citizen Kane*; Harold Urey (1893–1981), the American scientist who discovered heavy hydrogen ('heavy water') and was the recipient of the 1934 Nobel prize for chemistry; and Fred Zinnemann (1907–   ), the Austrian-born film director who was the recipient of Academy Awards for his films *From Here to Eternity* and *A Man for All Seasons*.

In 1945 the German army in Italy surrendered unconditionally to the Allies.

✎ 30 *April*

ON this day . . . in 1789 George Washington was inaugurated as the first US president.

In 1945 the German dictator Adolf Hitler and his mistress Eva Braun are believed to have committed suicide and had their bodies cremated in the bunker under the chancellory building in Berlin.

Born on this day were: Queen Mary II (1662–94), the eldest daughter of James II and wife of William III, who reigned jointly with her husband from 1689; Queen Juliana of The Netherlands (1909–), who abdicated on her birthday in 1980; and King Carl XVI Gustaf of Sweden (1946–), who succeeded his grandfather Gustavus VI in 1973.

Also born on this day were: Karl Friedrich Gauss (1777–1855), the German mathematician and astronomer who gave his name to the unit of magnetic flux density and (posthumously) the degaussing of magnetic mines in World War II; the Hungarian composer Franz Lehár (1870–1948), best known for his operetta *The Merry Widow*; and Jaroslav Hašek (1883–1923), the Czech author best known for his novel *The Good Soldier Schweik*.

&a 1 *May*

ON this day ... John Dryden, the first English Poet Laureate to be officially appointed to the post, died in 1700. (He was also the only Laureate to be dismissed, on refusing to take the oath to the new king, William III, after the Revolution in 1689.)

In 1945 Joseph Goebbels, the Nazi minister for propaganda, committed suicide with his wife and six children as Russian troops entered Berlin.

Born on this day were: the English essayist and poet Joseph Addison (1672–1719), the creator of 'Sir Roger de Coverley'; James Graham, the 6th Duke of Montrose (1878–1954), the first man to film a total eclipse of the sun and the inventor of the first naval aircraft-carrying ship; General Mark Wayne Clark (1896–1984), the commander of the US 5th Army in Italy and North Africa in World War II; and Joseph Heller (1923–), the American author who coined the expression 'Catch 22' in his novel of that name, based on his experiences as a serviceman in World War II.

&a 2 *May*

BORN on this day were: Harry Lillis Crosby (1901–77), the American singer and film actor who became famous as 'Bing' Crosby; and the British singer Arnold George Dorsey (1936–), who became famous as 'Engelbert Humperdinck'. (He took his name from the German composer.)

Also born on this day were: Alessandro Scarlatti (1660–1725), the Italian composer who was the father of the composer Domenico Scarlatti; Catherine II 'the Great' of Russia (1729–96), who succeeded her husband Peter III in 1762; Jerome K(lapka) Jerome (1859–1927), the English writer especially remembered for his humorous novel *Three Men in a Boat*; Theodor Herzl (1860–1904), the Hungarian-born writer who founded Zionism; Baron Manfred von Richthofen (1892–1918), the German fighter ace known as 'the Red Baron', who was credited with 'bringing down' a record eighty enemy aircraft in World War I; Dr Benjamin Spock (1903–), the American paediatrician who is credited with record sales of his *Commonsense Book of Baby and Child Care*; and the English composer Alan Rawsthorne (1903–71).

In 1960 Caryl Chessman, the author of the best-selling autobiography *Cell 2455, Death Row*, was executed in California's gas chamber, twelve years after being sentenced to death and in spite of world-wide protests.

3 May

BORN on this day were: Niccolò Machiavelli (1469–1527), the Italian statesman and political analyst whose name has become synonymous with unscrupulous pragmatism; Richard D'Oyly Carte (1844–1901), the London impresario who built the Savoy Theatre, where he produced the Gilbert and Sullivan operettas, the 'Savoy Operas'; the Scottish scientist John Scott Haldane (1860–1936); Gabriel Chevallier (1895–1969), the French author especially remembered for his novel *Clochemerle*; Dodie Smith (1896–1990), the English author especially remembered for her play *Autumn Crocus* and a children's book, *The Hundred and One Dalmatians*; Pete Seeger (1919–), the American folk singer and songwriter especially remembered for his song 'Where Have All the Flowers Gone'; the American boxer 'Sugar' Ray Robinson (1920–89), who won the world middleweight title for a record five times; and Henry Cooper (1934–), the British heavyweight champion who held the title for more than eleven years and was the first to win three Lonsdale Belts outright.

In 1951 George VI officially opened the Festival of Britain from the steps of St Paul's Cathedral, 100 years and 2 days after his great-grandmother Queen Victoria had opened the Great Exhibition of 1851.

✤ 4 May

ON this day ... in 1471 the Yorkists under Edward IV defeated the Lancastrians under Queen Margaret of Anjou, consort of Henry VI, at the battle of Tewkesbury, and Henry's son Prince Edward was killed.

The Derby was first run in 1780 on Epsom Downs, with Sir Charles Bunbury's Diomed in first place; the first British halfpenny newspaper, the *Daily Mail*, was published by Lord Northcliffe in 1896; and in 1979 Mrs Margaret Thatcher became Britain's first woman prime minister.

Born on this day were: Bartolommeo Cristofori (1655–1731), the Italian harpsichord maker who is credited with invention of the piano; Sir Thomas Lawrence (1769–1830), the English portrait painter; Joseph Whitaker (1820–95), the English bookseller and publisher who established (in 1868) *Whitaker's Almanack*; Thomas Henry Huxley (1825–95), the English biologist (and grandfather of Aldous and Julian Huxley) who coined the word 'agnostic'; John Hanning Speke (1827–64), the English explorer who discovered Lake Victoria and the source of the Nile; and the Belgian-born actress Audrey Hepburn (1929–93).

✤ 5 May

ON this day ... in 1961 the American astronaut Alan Shepard piloted *Mercury-Redstone III* in the first American sub-orbital space flight.

Born on this day were: Karl Marx (1818–83), the German philosopher and founder of modern international Communism; and the Danish philosopher Søren Aaby Kierkegaard (1813–55), the founder of modern existentialism.

Also born on this day were: John Batterson Stetson (1830–1906), the American hat manufacturer who gave his name to the wide-brimmed 'cowboy' hat; Nellie Bly (1867–1922), the American journalist and campaigner for women's rights; and Sir Gordon Richards (1904–86), the British champion jockey who rode his first (and last) Derby winner in 1953, at the age of 49.

In 1760 the fourth Earl Ferrers was driven from the Tower of London to Tyburn in his own landau to be hanged as a felon, the last English nobleman to be executed in this way and the first 'by the new drop just then introduced in the place of the barbarous cart, ladder, and mediaeval three-cornered gibbet'.

✍ 6 May

ON this day . . . in 1954 a British medical student, Roger Bannister, became the first man to run the mile in less than four minutes; and Robert Edwin Peary (1856–1920), the American admiral who was the first man to reach the North Pole (in 1909), was born on this day.

Also born on this day were: the French Revolutionary leader Maximilien Robespierre (1758–94), known as 'the Sea-green Incorruptible' (from his sallow complexion, recorded by the Scottish historian Thomas Carlyle); Sigmund Freud (1856–1939), the Austrian neurologist and founder of psychoanalysis (known as 'the Copernicus of the mind'); Sir Rabindranath Tagore (1861–1941), the Hindu poet and recipient of the 1913 Nobel prize for literature; Willem de Sitter (1872–1934), the Dutch scientist who calculated the size of the universe and developed the theory of the expanding universe; Sir Alan Cobham (1894–1973), the British aviation pioneer; and the Italian-born American film actor Rudolph Valentino (1895–1926), the celebrated 'screen lover' of the 1920s.

✍ 7 May

TWO great musicians were born on this day: the German composer Johannes Brahms (1833–97); and the Russian composer Peter Ilich Tchaikovsky (1840–93).

Also born on this day were: the English poet Robert Browning (1812–89); the British Liberal prime minister Lord Rosebery (1847–1929), who was the owner of three Derby winners; another Liberal politician, A(lfred) E(dward) W(oodley) Mason (1865–1948), better known as a novelist and the creator of 'Inspector Hanaud'; Josip Broz Tito, known as Marshal Tito (1892–1980), the Yugoslav leader from 1945 until his death; the American poet and politician Archibald MacLeish (1892–1982); Kitty Godfree (1896–1992), the English tennis player who was the Wimbledon ladies' singles champion in 1924 and 1926; the

American film actor Gary Cooper (1901–61); Edwin Herbert Land (1909–91), the American physicist who invented the 'Polaroid' camera; and the Australian-born champion jockey Arthur Edward 'Scobie' Breasley (1914–   ).

In 1915 an unarmed Cunard liner, the *Lusitania*, was torpedoed and sunk by a German submarine off Ireland, with the loss of almost 2,000 lives. (Among those who perished were 124 Americans and this is believed to have influenced the subsequent involvement of the USA in World War I.)

In 1945 all the remaining German forces surrendered unconditionally to the Allies.

✌ 8 *May*

ON this day . . . in 1945 the final surrender of the German armed forces the previous day was celebrated in Great Britain as VE (Victory in Europe) Day.

Harry S. Truman (1884–1972), the US president at that time, celebrated his 61st birthday. (The 'S.' in his name was inserted to conciliate his grandfathers, whose first names both began with that letter.)

Also born on this day were: Jean Henri Dunant (1828–1910), the Swiss philanthropist and founder of the Red Cross, who was the first recipient (jointly) of the Nobel peace prize, in 1901; the French film actor Fernand Constantin (1903–71), known as 'Fernandel'; the British naturalist and broadcaster Sir David Attenborough (1926–   ), whose brother is the actor Sir Richard Attenborough; and the American novelist Peter Benchley (1940–   ), whose father Nathaniel and grandfather Robert were also noted writers.

In 1794 the French scientist Antoine Lavoisier was guillotined for 'corruption'; the French fighter ace of World War I, Charles Nungesser, died in 1927, in an attempt to fly from Paris to New York (just thirteen days before Lindbergh's flight); and Lord 'Manny' Shinwell, the British Labour Party statesman, died in 1986 in his 102nd year.

❧ 9 *May*

BORN on this day were five famous people of the English theatre: the Scottish playwright Sir James Barrie (1860–1937), best remembered for his play *Peter Pan*; Lilian Baylis (1874–1937), the legendary manager of the Old Vic and Sadler's Wells Theatre; the playwright and actor Alan Bennett (1934–   ); the actor Albert Finney (1936–   ); and the actress Glenda Jackson (1936–   ), who in 1992 became a Labour Member of Parliament.

Also born on this day were: John Brown (1800–49), the American abolitionist whose attempts to abolish slavery by force played a part in starting the Civil War; and Howard Carter (1874–1939), the English archaeologist who with Lord Carnarvon discovered the tomb of the Egyptian King Tutankhamun, in 1922.

In 1926 the American aviators Richard Evelyn Byrd and Floyd Bennett made the first flight over the North Pole.

❧ 10 *May*

ON this day . . . in 1940 the British prime minister Neville Chamberlain resigned and was succeeded by Winston Churchill, at that time the First Lord of the Admiralty. ('I felt as if I were walking with destiny, and that all my past life had been but a preparation for this hour and this trial.') Exactly one year later Adolf Hitler's deputy Rudolf Hess parachuted from an aeroplane he had flown from Germany to Scotland and surrendered to the authorities with an Anglo-German 'peace plan'.

In 1863 Thomas Jonathan 'Stonewall' Jackson, the legendary Confederate general, died of wounds received at the battle of Chancellorsville on 2 May. (General Robert E. Lee, on hearing of his death, declared, 'I have lost my right arm.')

Born on this day were: Sir Thomas Lipton (1850–1931), the Scottish merchant and philanthropist who began work as an errand boy at the age of 9 and became a millionaire at 30 and was four times a challenger for the America's Cup with his yacht *Shamrock*; Léon Gaumont (1864–1946), the French motion picture pioneer; Dmitri Tiomkin (1894–1979), the Russian-born American composer and writer of numerous well-known film scores; and the American film actor, dancer, and singer Fred Astaire (1899–1987).

৯১ 11 *May*

ON this day... in 1812 the British prime minister Spencer Perceval was shot dead by a bankrupt broker, one John Bellingham, in the lobby of the House of Commons.

Born on this day were: Ottmar Mergenthaler (1854–99), the German inventor of the linotype typesetting machine; the legendary American jazz trumpeter Joe 'King' Oliver (1885–1938), mentor of Louis Armstrong; Irving Berlin (1888–1989), the Russian-born American songwriter who published the first of his 800 or more songs in 1907; Paul Nash (1889–1946), the English painter who was one of the first World War I official war artists on the Western Front; and the Spanish painter Salvador Dali (1904–89).

৯১ 12 *May*

ON this day... in 1641 Thomas Wentworth, the first Earl of Strafford, the former chief adviser to Charles I and known as 'Black Tom Tyrant', was beheaded (for treason) on Tower Hill by the executioner Richard Brandon, who was known as 'Young Gregory' from the first name of his father, his predecessor in the family 'business'. (Brandon also beheaded William Laud, the archbishop of Canterbury, and in 1649 the king himself, who had signed Strafford's death-warrant.)

Born on this day were: the German chemist Baron Justus von Liebig (1803–73), the discoverer of chloroform; Edward Lear (1812–88), the English artist and writer who popularized the limerick in his *Book of Nonsense*; Lord Grimthorpe (1816–1905), the English horologist who designed Big Ben; Florence Nightingale (1820–1910), the English nurse and hospital reformer who became famous for tending the sick and wounded soldiers in the Crimean War ('A lady with a Lamp shall stand | In the great history of the land...'—Henry Wadsworth Longfellow.); the English poet and painter Dante Gabriel Rossetti (1828–82), who was one of the founders of the Pre-Raphaelite Brotherhood; Leslie Charteris (1907–   ), the English writer of crime fiction and creator of 'the Saint'; and the English comedian Anthony John Hancock (1924–68), who created the comic character 'Tony Aloysius Hancock'.

※ 13 May

BORN on this day were: the Marquess of Rockingham (1730–82), the British prime minister 1765–6 and 1782; the French novelist Alphonse Daudet (1840–97); the English composer Sir Arthur Sullivan (1842–1900), best remembered for his collaboration with the playwright W. S. Gilbert; Sir Ronald Ross (1857–1932), the British physician best remembered for his discovery of the malarial parasite in the mosquito, for which he received the Nobel prize for physiology and medicine in 1902; the British painter Sir Frank Brangwyn (1867–1956); the French painter Georges Braque (1882–1963); the English novelist Dame Daphne du Maurier (1907–89); and Joe Louis (1914–81), the American world heavyweight boxing champion who held the title for a record eleven years.

※ 14 May

ON this day... in 1264 the English King Henry III was captured by his brother-in-law Simon de Montfort at the battle of Lewes; and in 1610 the French King Henri IV (Henri de Navarre) was assassinated by a religious fanatic, François Ravaillac.

Born on this day were: Gabriel Daniel Fahrenheit (1686–1736), the German physicist who invented and gave his name to the mercury thermometer; Thomas Wedgwood (1771–1805), the English physicist (and the son of Josiah Wedgwood) who is acknowledged as 'the first photographer' (though he was unable to find a way of 'fixing' the image); Robert Owen (1771–1858), the Welsh socialist and philanthropist who founded and gave his name to co-operative communities in Great Britain and the USA; the German aircraft designer Claude Dornier (1884–1969); the German-born conductor and composer Otto Klemperer (1885–1973); the American jazz musician Sidney Bechet (1897–1959), who pioneered the use of the soprano saxophone as a solo instrument in jazz (and who also died on this day); the English comedian Eric Morecambe (1926–84), who took his professional name from his home town in Lancashire; and Chay Blyth (1940–   ), the British yachtsman and circumnavigator who, in 1966, was the first to cross the Atlantic from America to Ireland by rowing boat (with Captain John Ridgway).

In 1940 the Home Guard ('Dad's Army') was formed as the LDV (Local Defence Volunteers), following a broadcast appeal by the British secretary of state for war Anthony Eden; and in 1948 Israel was proclaimed an independent state with the declaration that 'the State of Israel will be open to the immigration of Jews from all countries of their dispersion'.

ℰ 15 May

ON this day . . . in 1957 the first British hydrogen bomb was exploded near Christmas Island, in the Central Pacific; and in 1958 *Sputnik III*, the first 'space laboratory', was launched in the USSR.

Born on this day were: Lyman Frank Baum (1856–1919), the American journalist and playwright who created *The Wonderful Wizard of Oz*; Pierre Curie (1859–1906), the French scientist who with his wife Marie Curie discovered radium; the American actor Joseph Cotten (1905–   ); and the English actor James Mason (1909–84).

ℰ 16 May

ON this day . . . in 1763 James Boswell first met Dr Samuel Johnson. (Boswell: 'I do indeed come from Scotland, but I cannot help it . . .' Johnson: 'That, Sir, I find, is what a very great many of your countrymen cannot help.')

Born on this day were: Louis Vauquelin (1763–1829), the French scientist who discovered chromium; David Hughes (1831–1900), the London-born American inventor of the microphone; Sir Bernard Spilsbury (1877–1947), the British pathologist especially remembered for his conclusive evidence in the trial of 'Dr' Crippen; Richard Tauber (1891–1948), the Austrian-born tenor and composer of operettas; H(erbert) E(rnest) Bates (1905–74), the English novelist and short-story writer; the American film actor Henry Fonda (1905–82); and the American bandleader Woodrow Wilson 'Woody' Herman (1913–87).

In 1811 Wellington's General Beresford was victorious over Napoleon's Marshal Soult at the battle of Albuera, near Badajoz, in Spain.

✌ 17 *May*

ON this day . . . in 1900 the British forces commanded by General Baden-Powell at Mafeking were relieved after a siege of seven months; and in 1943 Wing Commander Guy Gibson led his 'Dam Busters' into the raid on the Mohne and Eder dams in the Ruhr, on the night of 17–18 May.

Born on this day were: Edward Jenner (1749–1823), the English physician who discovered vaccination; Timothy Healy (1855–1931), the first governor-general of the Irish Free State; the French composer Erik Satie (1866–1925): the French film actor Jean Gabin (1906–76); the second Viscount Maugham, the writer Robin Maugham (1916–81), who was a nephew of William Somerset Maugham; the Swedish soprano Birgit Nilsson (1918–   ); and the American boxing champion 'Sugar' Ray Leonard (1956–   ).

✌ 18 *May*

ON this day . . . in 1944 the battle of Monte Cassino ended as Allied troops finally captured the old fortified abbey (Europe's oldest monastic house), after more than three months of bombardment by shell-fire and air attack.

Born on this day were: William Steinitz (1836–1900), the Czech-born chess master and world champion in 1866–94, who became a naturalized American; and Fred Perry (1909–   ), the British-born tennis player and the first to win all four of the world's major titles (when he won the French title in 1935), who also became a naturalized American.

Also born on this day were: Lionel Lukin (1742–1834), the English coachbuilder who invented the first purpose-built 'insubmergible' lifeboat; Nicholas II (1868–1918), the last tsar of Russia; Bertrand Russell, the third Earl Russell (1872–1970), the English philosopher, mathematician, and recipient of the 1950 Nobel prize for literature; Walter Gropius (1883–1969), the German-born American pioneer of modern architecture; the American singer Perry Como (1912–   ); the French singer and songwriter Charles Trenet (1913–   ); and the Bulgarian-born bass singer Boris Christoff (1919–   ).

In 1812 John Bellingham, the assassin of the British prime minister Spencer Perceval, was hanged for murder, despite pleas of alleged insanity.

ॐ 19 *May*

BORN on this day were: Helen Porter Mitchell (1861–1931), who became famous as the prima donna Dame Nellie Melba, her professional name having been taken from Melbourne, in Australia, her place of birth; and Nancy Witcher Langhorne, Viscountess Astor (1879–1964), the first woman to sit in the House of Commons, who took her title from William Waldorf Astor, the second Viscount Astor, who had held her parliamentary seat until his elevation to the House of Lords. (Lord Astor was, coincidentally, also born on this day in 1879.)

Also born on this day were: the German philosopher Johann Gottlieb Fichte (1762–1814); the English architect Sir Albert Richardson (1880–1964), president of the Royal Academy 1954–7; Ho Chi Minh (1890–1969), the president of North Vietnam 1954–69; and the English film producer Sir Michael Balcon (1896–1977).

In 1536 Anne Boleyn was beheaded on Tower Green for adultery, on the eve of Henry VIII's marriage to Jane Seymour, his third wife and Anne's former lady-in-waiting. ('The executioner is, I believe, very expert; and my neck is very slender.')

In 1643 the young Duc d'Enghien led the French to victory against the 'invincible' Spanish army at the decisive battle of Rocroi, in the Ardennes.

ぞ 20 *May*

ON this day . . . in 1927 the American aviator Charles Lindbergh began his historic solo transatlantic flight, taking off from Roosevelt Field, New York, in his Ryan monoplane *Spirit of St Louis*; and in 1941 Crete was attacked by German paratroops (this being the first wholly airborne invasion of a country).

Born on this day were: the Earl of Shelburne, the British prime minister 1782–3; the French novelist Honoré de Balzac (1799–1850); the English philosopher and economist John Stuart Mill (1806–73); Emile Berliner (1851–1921), the German-born inventor of the gramophone using discs; Reginald Joseph Mitchell (1895–1937), the English aircraft designer who died too soon to witness the success of his fighter aircraft the Spitfire in World War II; Margery Allingham (1904–66), the English writer of crime fiction who created the amateur sleuth 'Albert Campion'; and the American film actor James Stewart (1908–   ).

Christopher Columbus died in poverty in 1506, still believing that he had discovered the coast of Asia; and the 'Northamptonshire peasant poet' John Clare died in the county lunatic asylum in 1864. ('If life had a second edition, how I would correct the proofs.')

ぞ 21 *May*

ON this day . . . in 1927 Charles Lindbergh landed at Le Bourget airfield, in Paris, to complete the first solo non-stop transatlantic flight.

Born on this day were: Albrecht Dürer (1471–1528), the German painter and engraver, generally recognized as the inventor of etching; Philip II of Spain (1527–98), whose second wife (he married four times) was Mary I of England; the English poet and satirist Alexander Pope (1688–1744); Elizabeth Fry (1780–1845), the English philanthropist and prison reformer; the American jazz pianist and composer Thomas 'Fats' Waller (1904–43); the American novelist Harold Robbins (1916–   ); and Andrei Sakharov (1921–89), the Russian physicist known as 'the father of the Soviet H-bomb', who was the recipient of the 1975 Nobel peace prize.

In 1471 Henry VI, the deposed king of England and the last of the Lancastrian kings, was murdered in the Tower of London.

✌ 22 May

On this day . . . in 1455 Henry VI was taken prisoner by the Yorkists at the first battle of St Albans, the first of the twelve battles of the English civil wars known as 'the Wars of the Roses' (the red rose being the
emblem of the Lancastrians and the white rose that of the Yorkists). In
1809, at the battle of Aspern-Essling, more than 40,000 men were killed
or wounded in an indecisive engagement between the armies of
Napoleon and Archduke Charles Louis of Austria.

Born on this day were: the German composer Richard Wagner
(1813–83); Sir Arthur Conan Doyle (1859–1930), the British author and
creator of the most celebrated fictional detective 'Sherlock Holmes'; and
the English actor Laurence Olivier (1907–89), who was knighted in 1947,
and made a life peer in 1970.

✌ 23 May

On this day . . . in 1498 Savonarola, the Dominican monk and
leader of the Democratic party in Florence, was hanged and burned for
heresy; in 1701 Captain William Kidd, the Scottish privateer and pirate,
was hanged on the banks of the Thames; in 1934 the American bank rob-
bers Bonnie Parker and Clyde Barrow (known popularly as 'Bonnie and
Clyde') were shot dead in a police ambush; and in 1945 the German
Gestapo chief Heinrich Himmler committed suicide after being captured by the British forces.

Born on this day were: Carolus Linnaeus (1707–78), the Swedish naturalist who gave his name to a system of classification for living things;
Franz Mesmer (1734–1815), the German physician who gave his name to
a kind of hypnotism; the English poet Thomas Hood (1799–1845); Otto
Lilienthal (1848–96), the German aviation pioneer who was killed flying
one of his own gliders; the English composer Edmund Rubbra
(1901–86); the American bandleader and clarinettist Artie Shaw
(1910–); the English bandleader and trumpet player Humphrey Lyt-
telton (1921–); Denis Compton (1918–), the English cricketer; the
American actress and singer Rosemary Clooney (1928–); and the
British-born American actress Joan Collins (1933–).

❧ 24 *May*

ON this day . . . in 1941 the battleship HMS *Hood*, at that time the largest capital ship in the world, was sunk by the German battleship *Bismarck*, with only three survivors from more than 400 on board.

Queen Victoria was born in 1819; and in 1895 John Henry Brodribb became the first actor to receive a knighthood (as Sir Henry Irving) in the Queen's Birthday Honours list.

Also born on this day were: William Gilbert (1540–1603), the royal physician to Elizabeth I and James I who gave his name to the unit of magneto-motive force; the English playwright Sir Arthur Wing Pinero (1855–1934); the South African soldier and statesman Field Marshal Jan Christiaan Smuts (1870–1950); Mikhail Sholokhov (1905–84), the Russian novelist and recipient of the 1965 Nobel prize for literature; the English playwright Arnold Wesker (1932–); and the American singer and songwriter Bob Dylan (1941–).

Nicolas Copernicus, the founder of modern astronomy, died in 1543, just after the printing of his book on the revolution of the celestial orbs, which he had delayed for some years for fear of official reaction.

❧ 25 *May*

ON this day . . . in 1935 the American athlete Jesse Owens set a record six world records in less than one hour at Ann Arbor, Michigan.

Born on this day were: Tom Sayers (1826–65), the English pugilist and middleweight champion who was only once beaten; and the American world heavyweight boxing champion Gene Tunney (1898–1978).

Also born on this day were: the Earl of Bute (1713–92), the first Scottish prime minister of Great Britain; the American poet and philosopher Ralph Waldo Emerson (1803–82); Pieter Zeeman (1865–1943), the Dutch physicist and recipient of the 1902 Nobel prize for physics who gave his name to the Zeeman effect, concerning optics; Igor Sikorsky (1889–1972), the Russian-born aeronautical engineer who designed the first successful helicopter used by the US forces in World War II; the American jazz trumpeter and composer Miles Davis (1926–91); and Sir Ian McKellen (1939–), the British actor.

✌ 26 *May*

ON this day . . . in 1865 the American Civil War ended with the surrender of the last Confederate army at Shrevport, near New Orleans.

The English diarist Samuel Pepys died in 1703. ('And so I betake myself to that course, which is almost as much as to see myself go into my grave; for which, and all the discomforts that will accompany my being blind, the good God prepare me.') The English poet Julian Grenfell was killed in action on the Western Front in 1915. ('And he is dead who will not fight; | And who dies fighting has increase.')

Born on this day were: Sir William Petty (1623–87), the author of the first book on vital statistics, *Political Arithmetic*; the French novelist Edmond de Goncourt (1822–96); the French aviation pioneer Henri Farman (1874–1958); Al Jolson (1886–1950), the American entertainer who starred in the first 'talkie', in 1927; the English composer Sir Eugene Goossens (1893–1962); John Wayne, the American actor (1907–79), whose name was originally Marion Michael Morrison; the English actor Robert Morley (1908–92); and the American singer Peggy Lee (1920–   ).

✌ 27 *May*

ON this day . . . in 1941 the *Bismarck*, the largest and most powerful German battleship of its time, was sunk by ships of the Royal Navy, just three days after her sinking of the British battleship HMS *Hood*.

Born on this day were: Mrs Amelia Jenks Bloomer (1818–94), the American campaigner for women's suffrage who popularized and gave the name to 'bloomers'; another American suffragist, Mrs Julia Ward Howe (1819–1910), who wrote 'The Battle Hymn of the Republic' ('Mine eyes have seen the glory . . .'); James Butler Hickok (1837–76), the US marshal immortalized as 'Wild Bill Hickok'; the English novelist (Enoch) Arnold Bennett (1867–1931), whose novels immortalized the 'five towns' in his native Staffordshire; the French painter Georges Rouault (1871–1958); the French composer Louis Durey (1888–1979), the senior member of 'Les Six'; Dashiell Hammett (1894–1961), the American writer of detective fiction who was formerly a 'Pinkerton' man; Herman Wouk (1915–   ), the American writer and winner of the

Pulitzer prize (in 1951) for his novel *The Caine Mutiny*; and Dr Henry Kissinger (1923–), the former US secretary of state and joint recipient of the 1973 Nobel peace prize.

೫ 28 *May*

BORN on this day was William Pitt (1759–1806), the youngest British prime minister (in 1783, aged 24) and the first to follow his father into the premiership.

In 1959 the Mermaid Theatre, the first new theatre in the City of London since the time of Shakespeare, was opened by the actor Bernard Miles (later Lord Miles).

Also born on this day were: Joseph Guillotin (1738–1814), the French physician who advocated the employment of the beheading machine to which he gave his name, but did not invent; the Irish poet and musician Thomas Moore (1779–1852); the English novelist Warwick Deeping (1877–1950); Sir George Dyson (1883–1964), the English composer who was (unusually) the author of the official War Office publication *The Manual of Grenade Fighting*, in World War I; the English novelist Ian Fleming (1908–64), who created the secret service agent 'James Bond'; the Australian novelist Patrick White (1912–), the recipient of the 1973 Nobel prize for literature; and the German baritone and *Lieder* singer Dietrich Fischer-Dieskau (1925–).

೫ 29 *May*

ON this day ... in 1453 the Byzantine Empire came to an end with the capture of Constantinople by the Turks, who made it their capital.

In 1660 Oak Apple Day was first celebrated on Charles II's return to London on his 30th birthday, following his Restoration to the throne. (Oak leaves had been worn in commemoration of the king's escape after the battle of Worcester in 1651, when he hid in an oak tree at Boscobel.)

In 1953 the New Zealander Edmund Hillary and Sherpa Tenzing Norgay (1914–86) became the first men to climb to the summit of Mount

Everest (and the news of their achievement reached England four days later, on the day of the coronation of Queen Elizabeth II).

Born on this day (in addition to Charles II) were: the Spanish composer Isaac Albéniz (1860–1909); the Austrian composer Erich Korngold (1897–1957), who became especially well known for his film music; G(ilbert) K(eith) Chesterton (1874–1936), the English poet and novelist and creator of the detective 'Father Brown'; the British-born American comedian and film actor Bob Hope (1903–   ); Beatrice Lillie (1898–1989), the Canadian-born actress who became Lady Peel when she married Sir Robert Peel, a descendant of the British prime minister; and John Fitzgerald Kennedy (1917–63), the thirty-fifth (and youngest) US president.

ℬ 30 *May*

ON this day . . . in 1431 the French patriot and martyr Joan of Arc was burned at the stake at Rouen, after being captured by the English. (She was canonized in 1920.)

The English dramatist Christopher Marlowe was killed in a tavern brawl in London, in 1593.

Born on this day were: Alfred Austin (1835–1913), the English Poet Laureate who succeeded Tennyson some four years after his death; and the American clarinet virtuoso and bandleader Benny Goodman (1909–86), known as 'the king of swing'.

In 1942 the first British 'thousand bomber raid' took place on Cologne. (And a BBC broadcast of the cellist Beatrice Harrison, playing to the accompaniment of nightingales in a Surrey wood, was cancelled at the last moment to prevent the noise of the aircraft giving an early warning to the enemy.)

ℬ 31 *May*

ON this day . . . in 1902 the signing of the Treaty of Vereeniging ended the Boer War; and in 1916 the battle of Jutland, the major naval battle of World War I, was fought between the British fleet under Jellicoe and Beatty and the German fleet under Von Scheer. (The British lost 14 of their 149 ships and the Germans 11 of their 110 in the engagement, which was indecisive; though the German fleet was deterred from giving open battle a second time.)

Born on this day were: the American poet Walt(er) Whitman (1819–92); the English artist Walter Sickert (1860–1942); the English artist William Heath Robinson (1872–1944), whose caricatures of mechanical contraptions gave his name to the English language; the British actress and centenarian Athene Seyler (1889–1990); the American actor Clint Eastwood (1930–   ); and Terry Waite (1939–   ), the former adviser to the archbishop of Canterbury who was held as a hostage in the Middle East for almost five years from 1987.

In 1962 Adolf Eichmann, the former SS officer, was hanged near Tel Aviv for his part in the murder of more than a million Jews. (This was the first execution to take place in the State of Israel.)

JUNE ✃

✃ 1 June

ON this day ... in 1794 the French fleet under Admiral Villaret de Joyeuse was defeated by the British under Lord Howe in the second battle of Ushant, known as the battle of 'the Glorious First of June'.

In 1831 the Scottish explorer James Clark Ross located the North Magnetic Pole.

Born on this day were: Henry Francis Lyte (1793–1847), the English clergyman who wrote the hymn 'Abide with me', a few weeks before his death; Brigham Young (1801–77), the American Mormon leader and founder of Salt Lake City; Mikhail Ivanovich Glinka (1804–57), the Russian composer whose birthplace Novospasskoye was renamed Glinka in his honour; John Masefield (1878–1967), the English poet, playwright, and novelist and Poet Laureate from 1930; the English poet and playwright John Drinkwater (1882–1937); the English playwright John Van Druten (1901–57); Air Commodore Sir Frank Whittle (1907–), a pioneer in the development of the jet engine; and the American film actress and sex symbol Marilyn Monroe (1926–62).

In 1943 the internationally famous British actor Leslie Howard died when an unescorted passenger aircraft in which he was returning to England from Spain was deliberately attacked and destroyed by German war-planes. (It has been suggested that German intelligence believed that the British prime minister Winston Churchill was also on the flight.)

✃ 2 June

BORN on this day were: the English novelist and poet Thomas Hardy (1840–1928); and the English composer Sir Edward Elgar (1857–1934).

In 1953, the coronation of HM Queen Elizabeth II took place in Westminster Abbey; and during that day the news arrived of the successful ascent of Mount Everest by the New Zealander Edmund Hillary and the Sherpa Tenzing Norgay of the British expedition led by Colonel John Hunt.

Also born on this day were: Comte Donatien Alphonse François de Sade (1740–1814), the French writer known as the Marquis de Sade, from whose name and sexual perversions we have the word 'sadism'; Jesse Boot, first Baron Trent of Nottingham (1850–1931), who founded Boot's the chemists; the Austrian conductor and composer Felix Weingartner (1863–1942); Johnny Weissmuller (1904–84), the Olympic swimming champion who became the best known 'Tarzan' in films; Johnny Speight (1920–   ), the British scriptwriter and creator of 'Alf Garnett'; and Marvin Hamlisch (1944–   ), the American composer who created the music for *A Chorus Line*.

In 1989 more than 100 students and other civilians were killed when Chinese soldiers opened fire on a non-violent protest in Tiananmen Square, in Peking.

✌ 3 June

On this day . . . in 1940 the evacuation of the Allied forces from Dunkirk was completed; and in 1942 the battle of Midway began as the Japanese engaged the US Navy in one of the decisive conflicts of World War II.

The French composer Georges Bizet died in 1875, aged 36, believing that his opera *Carmen* was a failure; and the Austrian novelist Franz Kafka died in 1924, aged 40, with none of his novels having been published.

Born on this day were: the Reverend Sydney Smith (1771–1845), the English essayist, wit, and epicure; František Jan Škroup (1801–62), the composer of the Czech national anthem; Richard Cobden (1804–65), the English economist and politician known as 'the Apostle of Free Trade'; Jefferson Davis (1808–89), the president of the Confederate States of America in the Civil War; the French composer Charles Lecocq (1832–1918); the French painter Raoul Dufy (1877–1953); Josephine Baker (1906–75), the American singer and dancer; the dramatist William Douglas-Home (1912–92), who is the brother of Lord Home, the former British prime minister; the American film actor Tony Curtis (1925–   ); and the American poet Alan Ginsberg (1926–   ).

In 1937 the Duke of Windsor, who as Edward VIII had abdicated the previous year, was married to the American divorcee Mrs Simpson, on the anniversary of the birth of his father George V in 1865.

ᴋᴀ 4 June

On this day . . . in 1844 the last known specimen of the garefowl, the great auk, was killed on the Stack of Eldey, off SW Iceland.

In 1913 a young suffragette, Emily Davison, threw herself under the king's horse Anmer at the Epsom Derby.

Born on this day were: George III (1738–1820), whose reign was the longest of any king of Great Britain; Mary St Leger Kingsley (1852–1931), the English novelist who wrote under the pseudonym of 'Lucas Malet' and was the daughter of the writer Charles Kingsley; Carl Gustaf Emil von Mannerheim (1867–1951), the Finnish soldier and statesman who gave his name to a line of defence against Russia; and Sir Christopher Cockerell (1910–), the inventor of the hovercraft.

In 1859 the French under Napoleon III defeated the Austrians at the battle of Magenta in the Austro-Sardinian War. (Neither artillery nor cavalry were employed in the battle.)

ᴋᴀ 5 June

On this day . . . in 1783 the first ascent by a hot-air balloon, built by the French brothers Montgolfier, lasted ten minutes.

In 1916 the cruiser HMS *Hampshire* struck a mine off Orkney and sank, with the secretary of state for war Lord Kitchener among those who perished; and in 1968 the US senator Robert Kennedy, brother of the former president John F. Kennedy, was shot by a Jordanian-born immigrant in Los Angeles, as he was campaigning for the Democratic presidential nomination. (He died the next day.)

Born on this day were: the Scottish economist Adam Smith (1723–90), who wrote the first serious work on political economy, *An Enquiry into the Nature and Causes of the Wealth of Nations*; and John Maynard Keynes (1883–1946), the English economist whose 'Keynesian' theory regarding full employment through state control of the economy had a world-wide influence.

Also born on this day were: the English novelist Dame Ivy Compton-Burnett (1884–1969); Dr Kurt Hahn (1886–1974), the German-born founder of Gordonstoun School, whose pupils included Prince Philip and the Prince of Wales; the Spanish poet and dramatist Federico García Lorca (1899–1936), who was murdered by Franco's supporters at the start of the Civil War; and the English novelist and biographer Margaret Drabble (1939–   ), editor of *The Oxford Companion to English Literature*.

In 1967 a lightning attack by Israel on Arab territory signalled the start of 'the Six-Day War' in the Middle East.

๕๑ 6 June

ON this day . . . in 1944 the invasion of Nazi-occupied Europe began as Allied forces landed in Normandy in the greatest sea and air operation in military history, code-named 'Overlord' but better known as D-Day.

Born on this day were: the French dramatist Pierre Corneille (1606–84), best remembered for *Le Cid*; the English composer Sir John Stainer (1840–1901), best remembered for his oratorio *The Crucifixion*; William Ralph Inge (1860–1954), the English scholar and dean of St Paul's known as 'the Gloomy Dean' from his pessimistic oratory; Sir Henry Newbolt (1862–1938), the English poet and official naval historian; Captain Robert Falcon Scott (1868–1912), who with four others was the first Englishman to reach the South Pole; the German writer Thomas Mann (1875–1955); the English writer R(obert) C(edric) Sherriff (1896–1975); the British ballerina and choreographer and founder of the Royal Ballet, Dame Ninette de Valois (1898–   ); the Armenian composer Aram Khachaturyan (1903–78); and Bjorn Borg (1956–   ), the Swedish tennis player who won a record five consecutive Wimbledon men's singles championships in 1976–80.

✌ 7 June

BORN on this day were: the French painter Paul Gauguin (1848–1903) whose bizarre life story (he abandoned his family and regular way of living to devote himself to painting in Tahiti) was the subject of Somerset Maugham's novel *The Moon and Sixpence*; and the Italian painter Pietro Annigoni (1910–88), who abandoned portrait painting of the famous (including the British royal family) for the traditional methods of the old masters.

Also born on this day were: John Rennie (1761–1821), the Scottish engineer who designed many of London's bridges; the Earl of Liverpool (1770–1828), the British prime minister 1812–27, who came to office following the assassination of Spencer Perceval; George Bryan Brummell, known as 'Beau' Brummell (1778–1840), the leader of fashion and associate of the Prince Regent; the English novelist Richard Doddridge Blackmore (1825–1900), who created the eponymous heroine 'Lorna Doone'; Ernest William Hornung (1866–1921), the English novelist who created the gentleman burglar 'Raffles'; and John Cameron Andrieu Bingham Morton, known as J. B. Morton or 'Beachcomber' (1893–1979), the English journalist and satirical writer.

In 1915 the Royal Naval Air Service pilot Flight sub-Lieutenant R. A. J. Warneford became the first airman to destroy a Zeppelin (over Belgium), for which he was awarded the first RNAS Victoria Cross.

✌ 8 June

ON this day ... in 1929 Margaret Bondfield became Britain's first woman cabinet minister with her appointment as minister of labour in Ramsay MacDonald's second Labour government; and in 1978 (Dame) Naomi James completed the first solo circumnavigation of the world by a woman. Born on this day were: the English engineer John Smeaton (1724–92), who constructed the third Eddystone lighthouse, in 1759; and Robert Stevenson (1772–1850), the Scottish engineer (and grandfather of Robert Louis Stevenson) who constructed the Bell Rock lighthouse.

Also born on this day were: the German composer Robert Schumann (1810–56); Charles Reade (1814–84), the English novelist and dramatist best known for his historical novel *The Cloister and the Hearth*; Sir John Everett Millais (1829–96), the English painter and one of the originators of the Pre-Raphaelite Brotherhood; the American architect Frank Lloyd Wright (1869–1959); and the British scientist Francis Crick (1916–   ), joint recipient of the 1962 Nobel prize for medicine and physiology for the discovery of the structure of DNA (the carrier of genetic information).

In 1405 Richard Scrope, archbishop of York, was beheaded for conspiring against Henry IV. (He was the first English bishop to be executed.)

9 *June*

ON this day . . . in 1931 the Davis escape apparatus was first used in an emergency when six men escaped from HM Submarine *Poseidon*, which had sunk off China; and in 1959 the first submarine armed with ballistic missiles, *The George Washington*, was launched.

Born on this day were: George Stephenson (1781–1848), the English inventor of the first locomotive for a public railway, the Stockton and Darlington Railway (in 1825), and the famous *Rocket* for his Liverpool and Manchester Railway; Otto Nicolai (1810–49), the German composer best known for his overture *The Merry Wives of Windsor*; Elizabeth Garrett Anderson (1836–1917), the first British woman to qualify as a doctor and the first woman mayor in Britain; the Danish composer Carl Nielsen (1865–1931); and the American composer and songwriter Cole Porter (1891–1964).

Charles Dickens died in 1870 and was later buried in Westminster Abbey.

10 *June*

ON this day . . . in 1934 the English composer Frederick Delius died in France, having been paralysed and blind for the last years of his life, yet still composing with his amanuensis Eric Fenby; and in 1944 German soldiers shot all the men of the French town of Oradour-sur-Glane and drove the women and children into the church which was then set on fire. (The town has since remained unoccupied as a memorial to this atrocity, which was apparently a reprisal for the D-Day landings.)

Born on this day were: James Francis Edward Stuart, 'the Old Pretender' (1688–1766), the only son of James II by Mary of Modena; John Dollond (1706–61), the English optician and inventor of telescopes; the French artist Gustave Courbet (1819–77); the French artist André Derain (1880–1954); Eric Maschwitz (1901–69), the English playwright and songwriter (best remembered for 'These Foolish Things', 'A Nightingale Sang in Berkeley Square', and 'Room 504'); the American song composer Frederick Loewe (1901–88); Sir Terence Rattigan (1911–77), the English playwright; Saul Bellow (1915–   ), the American novelist and recipient of the 1976 Nobel prize for literature; HRH Prince Philip, Duke of Edinburgh (1921–   ); Judy Garland (1922–69), the American singer and film actress; and Robert Maxwell (1923–91), the British publisher and newspaper proprietor.

In 1829 the first Oxford and Cambridge boat race, from Hambledon Lock to Henley Bridge, was won by Oxford in 14 minutes and 30 seconds; and in 1909 an SOS signal was transmitted in an emergency for the first time, as the Cunard liner SS *Slavonia* was wrecked off the Azores.

ൠ 11 *June*

On this day . . . in 1488 James III of Scotland was murdered by rebellious Scottish nobles after the battle of Sauchieburn and was succeeded by his 15-year-old son James IV.

Born on this day were: John Constable (1776–1837), the British landscape painter; Richard Strauss (1864–1949), the German composer; Nikolai Bulganin (1895–1975), the Soviet premier 1955–8; Jacques Cousteau (1910–   ), the French marine explorer and joint inventor of the aqualung; the British actor Richard Todd (1919–   ); Dame Beryl Grey (1927–   ), the prima ballerina; and Jackie Stewart (1939–   ), the Scottish racing driver and world champion in 1969, 1971, and 1973.

In 1847 the Arctic explorer Sir John Franklin died in an attempt to discover the North-West Passage, the route to the East round the north of the American continent.

12 *June*

BORN on this day were: Sir Anthony Eden, later first Earl of Avon (1897–1977), the British prime minister 1955–7; and George Bush (1924–), the forty-first US president.

Also born on this day were: Charles Kingsley (1819–75), the English clergyman and Christian Socialist especially remembered for his allegorical children's book *The Water Babies*; Rikard Nordraak (1842–66), the Norwegian remembered as the composer of the Norwegian national anthem; Leon Goossens (1897–1988), the oboist and one of the famous Goossens family of musicians; and Anne Frank (1929–45), the Jewish girl whose now famous diary was published in 1947, two years after her death in the Belsen concentration camp.

In 1667 a 15-year-old French boy became the first human being to be given a blood transfusion when the personal physician to Louis XIV, Jean-Baptiste Denys, successfully used lamb's blood to cure his fever.

13 *June*

THE Irish poet and playwright William Butler Yeats (1865–1939), the recipient of the Nobel prize for literature in 1923, was born on this day.

In 1944 the first flying bomb, the VI (Vergeltungswaffe 1), known as the 'doodlebug', was first used by the Germans against Britain.

Also born on this day were: Fanny Burney, Madame d'Arblay (1752–1840), the English novelist who was the daughter of the music historian Charles Burney; Dr Thomas Arnold (1795–1842), the headmaster of Rugby, immortalized in Thomas Hughes's novel *Tom Brown's Schooldays*, who was the father of the poet Matthew Arnold; the German soprano Elisabeth Schumann (1888–1952); the South African-born film actor Basil Rathbone (1892–1967) who became the definitive 'Sherlock Holmes' of the screen; Dorothy L(eigh) Sayers (1893–1957), the English writer who created another famous amateur detective, 'Lord Peter Wimsey'; and John Donald Budge (1915–), the American tennis player who was the first to hold all four of the major championships simultaneously, in 1938.

℘ 14 *June*

On this day ... in 1645 the royalists under Charles I and Prince Rupert were defeated by the parliamentarians under Cromwell and Fairfax at Naseby, in Northamptonshire, in what was probably the decisive battle of the Civil War and the first major engagement of the New Model Army.

Born on this day were: Harriet Beecher Stowe (1811–96), the American author whose anti-slavery novel *Uncle Tom's Cabin* is said to have played a part in the commencement of the American Civil War; Count John McCormack (1884–1945), the Irish-born naturalized American tenor; the English tenor Heddle Nash (1896–1961); the English poet Kathleen Raine (1908–   ); the American singer and film actor Burl Ives (1909–   ); and the German tennis champion Steffi Graf (1969–   ).

Sir Harry Vane, a Civil War leader on the parliamentary side (though not a regicide), was beheaded for treason in 1662.

In 1800 the French general Louis Charles Antoine Desaix was killed by a musket ball at the battle of Marengo, after saving the day for Napoleon in one of the major engagements of the Italian campaign.

℘ 15 *June*

On this day ... in 1215 the Magna Carta (the Great Charter) was signed by King John at Runnymede, in Surrey, in the presence of the English barons, establishing the rights of political and personal liberty; in 1381 the leader of the Peasants' Revolt, Wat Tyler, was struck down by William Walworth, mayor of London, in the process of negotiations with the young King Richard II, and was beheaded shortly afterwards; and in 1977 the first general election in Spain since 1936 resulted in victory for the UCD (Union of Democratic Centre).

In 1919, the British aviators Alcock and Brown completed the first non-stop transatlantic flight from Newfoundland to Ireland in a Vickers Vimy bomber plane. (The aircraft has been preserved in the Science Museum in London.)

Born on this day were: Edward, the Black Prince (1330–76), the eldest son of Edward III, the first duke created in England (as the Duke of

Cornwall); Josiah Henson (1789–1883), the American negro who escaped from slavery and became an internationally famous preacher and the reputed prototype of 'Uncle Tom' in Harriet Beecher Stowe's celebrated novel; Hablot Knight Browne (1815–82), the artist and illustrator known as 'Phiz', who illustrated *Pickwick Papers* and other works by Dickens; and the Norwegian composer Edvard Grieg (1843–1907), best known for his music to Ibsen's *Peer Gynt*.

✍ 16 June

ON this day . . . in 1963 the Russian astronaut Valentina Tereshkova became the first woman to be launched into space.

Born on this day were: Sir John Cheke (1514–57), the English classical scholar who was the tutor of Edward VI and secretary of state for Lady Jane Grey (and who was celebrated in verse by John Milton: 'Thy age, like ours, O soul of Sir John Cheke, | Hated not learning less than toad or asp, | When thou taught'st Cambridge and King Edward Greek.'); and John Enoch Powell (1912–), the English classical scholar and former Conservative minister.

Also born on this day were: Julius Plucker (1801–68), the German physicist who discovered cathode rays; Sir George Frampton (1860–1928), the British sculptor whose works include the *Peter Pan* statue in Kensington Gardens and the Edith Cavell memorial near the Church of St Martin-in-the-Fields; and Tom Graveney (1927–), the England cricketer and scorer of 122 First Class centuries.

In 1815 Wellington's army defeated Marshal Ney's forces at Quatre-Bras, the battle preceding Waterloo.

An element of liberty Magna Carta

✍ 17 June

ON this day . . . in 1775 the battle of Bunker Hill, the first major engagement of the American War of Independence, resulted in victory for the English but at great cost in lives and the morale of the troops.

Born on this day were: Edward I (1239–1307), king of England from the death of his father Henry III in 1272; John Wesley (1703–91), the English evangelist and founder of Methodism; William Parsons, the third Earl of Rosse (1800–67), the astronomer who built the world's largest telescope (in 1845); the French composer Charles Gounod (1818–93), who sold his famous opera *Faust* to his English publishers for £80; the Russian-born composer Igor Stravinsky (1882–1971), who mistakenly adjusted his birthday date from the Old Style (5 June) to 18 June, on becoming an American citizen; the American singer and film actor Dean Martin (1917–   ); and the American singer and songwriter Barry Manilow (1946–   ).

In 1972 five men were arrested by Washington police in the act of breaking into the Democratic Party National Committee's headquarters, the Watergate building, the incident which initiated the 'Watergate Affair' and the subsequent resignation of President Nixon in 1974.

৯৯ 18 *June*

ON this day . . . in 1815 Napoleon was finally defeated by Wellington and Blücher at Waterloo, near Brussels. ('Nothing except a battle lost can be half so melancholy as a battle won.')

In 1916 Max Immelmann, the World War I German fighter pilot who originated and gave his name to the manœuvre the Immelmann turn, died in action; and in 1928 the American airwoman Amelia Earhart became the first woman to fly the Atlantic.

Born on this day were: Robert Stewart, Viscount Castlereagh (1769–1822), who as Foreign Secretary represented England at the Congresses of Châtillon and Vienna in 1814–15, and at the Treaty of Paris in 1815 (following the close of Napoleon's last campaign) and the Congress of Aix-la-Chapelle; Édouard Daladier (1884–1970), the French prime minister who represented his country at the Munich Conference with Chamberlain, Hitler, and Mussolini; the American songwriter and entertainer Sammy Cahn (1913–93); the British actor Ian Carmichael (1920–   ); the British actor Paul Eddington (1927–   ); and Paul McCartney (1942–   ), the British musician and songwriter and one of the original 'Beatles'.

☙ 19 *June*

On this day . . . in 1867 Ferdinand Joseph Maximilian, archduke of Austria and emperor of Mexico from 1864, was condemned to death and shot by his opponents; and in 1953 Julius Rosenberg and his wife Ethel were executed in the electric chair at Sing Sing prison, New York, for espionage, despite world-wide appeals for clemency.

Born on this day were: the French scientist and philosopher Blaise Pascal (1623–62), who gave his name to a unit of pressure and invented one of the earliest calculators; and Sir Ernst Boris Chain (1906–79), the German-born scientist who shared the 1945 Nobel prize for medicine with Sir Alexander Fleming and Sir Howard Florey for his work on the development of penicillin.

Also born on this day were: James I, king of Great Britain and formerly James VI of Scotland (1566–1625); Douglas Haig, First Earl Haig of Bemersyde (1861–1928), commander-in-chief of the British forces in France 1915–19; Sir William Ashbee Tritton (1875–1946), the designer of the first tank to go into action and named after him as 'Big Willie'; Walter Reginald 'Wally' Hammond (1903–65), the England cricketer and scorer of 167 First-Class centuries; and the Indian-born novelist Salman Rushdie (1947–   ), Booker prizewinner in 1981 and author of the controversial *Satanic Verses*.

☙ 20 *June*   Boxer rebellion 1900

On this day . . . in 1756, during the Anglo-French struggle for India, 145 men and one woman from the captured British garrison at Calcutta were incarcerated in a cell less than eighteen feet square, by order of Suraja Dowlah, the Nawab of Bengal, and only twenty-three survived the night. (This came to be known as 'the Black Hole of Calcutta Atrocity'.) In 1789 the first significant act in the French Revolution took place, when the National Assembly were ordered out of their usual meeting-place and assembled in a nearby tennis court, where an oath was taken—never to disperse until they had given France a constitution. (This came to be known as 'the Oath of the Tennis Court'.)

Born on this day were: Jacques Offenbach (1819–80), the German-born French composer; Sir Frederick Gowland Hopkins (1861–1947), the English biochemist and joint recipient of the 1929 Nobel prize for physiology and medicine for the discovery of vitamins; Lillian Hellman (1905–84), the American playwright and screenwriter; Dame Catherine Cookson (1906–   ), the English novelist; the Australian-born film actor Errol Flynn (1909–59); and the film actor Audie Murphy (1924–71), who was the most decorated American soldier of World War II.

☙ 21 *June*

ON this day . . . in 1854 the first Victoria Cross was won by Mate Charles David Lucas (later Rear-Admiral Lucas) on board HMS *Hecla*, when he threw a live shell overboard a few seconds before it exploded; and in 1919 Admiral Reuter gave the order to scuttle the interned German fleet at Scapa Flow.

Born on this day were: Charles Edward Horn (1786–1849), the English composer and singer best known for his song 'Cherry Ripe'; Joseph Kesselring (1902–67), the American playwright best known for *Arsenic and Old Lace*; Jean-Paul Sartre (1905–80), the French existentialist philosopher, dramatist, and novelist who was awarded, but declined to accept, the 1964 Nobel prize for literature; the American novelist Mary McCarthy (1912–89); the American film actress Jane Russell (1921–   ), sex symbol of the post-war decade; the French novelist Françoise Sagan (1935–   ); and Benazir Bhutto (1953–   ), the former prime minister of Pakistan whose father, the previous prime minister, was executed in 1979.

In 1813 Wellington defeated Napoleon's brother Joseph at the battle of Vitoria, the major battle of the Peninsular War.

☙ 22 *June*

ON this day . . . in 1679 the rebellion of the Scottish Covenanters, the strict Presbyterian sect, was put down by the Duke of Monmouth at the battle of Bothwell Bridge; and in 1941 the German invasion of the USSR began.

Born on this day were: the English novelist Sir Henry Rider Haggard (1856–1925), best remembered for *King Solomon's Mines*; and the German novelist Erich Maria Remarque (1898–1970), best remembered for *All Quiet on the Western Front.*

Also born on this day were: Theodor Leschetizky (1830–1915), the Polish pianist and a renowned teacher of the piano, whose pupils included Paderewski and Schnabel (and whose second, third, and fourth wives were also his pupils); Sir Julian Huxley (1887–1975), the English biologist who was a grandson of T. H. Huxley and a brother of Aldous Huxley; the Austrian-born film director Billy Wilder (1906–   ); the American showman and film producer Mike Todd (1909–58); the English tenor Sir Peter Pears (1910–86), for whom Benjamin Britten wrote all his major tenor roles; and the American actress Meryl Streep (1949–   ).

In 1535 John Fisher, bishop of Rochester, was beheaded on Tower Hill for refusing to swear to the royal supremacy of Henry VIII, shortly after the pope had made him a cardinal. (He was canonized in 1935.)

✌ 23 *June*

On this day . . . in 1757 the army of the Nawab of Bengal, who had perpetrated the 'Black Hole of Calcutta Atrocity' the year before, was defeated by a small force led by Robert Clive at the battle of Plassey.

Born on this day were: John Fell (1625–86), the bishop of Oxford who was the subject of the well-known verse 'I do not love thee, Dr Fell . . .'; the Empress Josephine (1763–1814), Napoleon's first wife; Edward VIII (1894–1972), the king of Great Britain who became the Duke of Windsor after his abdication in 1936; Dr Alfred Kinsey (1894–1956), the American sexologist and author of the controversial (at the time) *Sexual Behavior in the Human Male*; the French dramatist Jean Anouilh (1910–87); William Pierce Rogers (1913–   ), the US secretary of state 1969–73; and Sir Leonard Hutton (1916–90), the England cricketer who was the first professional player to captain his country at cricket.

❧ 24 June

ON this day . . . in 1314 Robert Bruce defeated Edward II's army at the battle of Bannockburn, securing independence for Scotland; and in 1859 a battle between the French and the Austrians at Solferino, in northern Italy, led to the establishment of the International Red Cross by the Swiss philanthropist Jean Henri Dunant, who witnessed the suffering of the wounded on the battlefield.

Two famous British soldiers were born on this day: John Churchill, first Duke of Marlborough (1650–1722); and Horatio Herbert Kitchener, first Earl Kitchener of Khartoum (1850–1916).

Also born on this day were: William Henry Smith (1825–91), the First Lord of the Admiralty (1877–80) in Disraeli's cabinet, but better remembered for his work in expanding his father's bookselling business, W. H. Smith; Ambrose Bierce (1842–1913?), the American journalist and short-story writer, whose date of death is unknown since he disappeared in Mexico in 1913; Jack Dempsey (1895–1983), the American world heavyweight boxing champion 1919–26; Juan Fangio (1911–   ), the Argentinian racing motorist and five times a world champion; and Brian Johnston (1912–   ), the British broadcaster and writer (especially on cricket).

❧ 25 June

ON this day . . . in 1876 Colonel George Armstrong Custer and all his 264 men were killed by the Sioux Indians at the battle of Little Big Horn, which has become known as 'Custer's Last Stand'.

In 1483 Earl Rivers and Lord Richard Grey, the uncle and stepbrother of the young Edward V, were executed by order of Richard III, who on the same day had deposed his nephew.

Born on this day were: the French composer Gustave Charpentier (1860–1956); Erskine Childers (1870–1922), the Irish writer who was executed by the Irish Free State authorities; Lord Mountbatten of Burma (1900–79), the British admiral and statesman who was assassinated by Irish terrorists; and the English novelist George Orwell (1903–50), whose real name was Eric Blair.

In 1950 the Korean War began as North Korean troops crossed the 38th parallel.

ᏚᎣ 26 *June*

ON this day ... in 1483 Richard, Duke of York and Protector of the young Edward V, assumed the crown as Richard III; and in 1830 William IV began his reign with the death of his brother George IV.

Born on this day were: George Morland (1763–1804), the English painter who exhibited at the Royal Academy at the age of 10; Patrick Branwell Brontë (1817–48) the profligate brother of the Brontë sisters and 'black sheep' of the famous family; William Thomson, 1st Baron Kelvin (1824–1907), the British scientist who gave his name to the Kelvin scale of temperature; George Edward Stanhope Molyneux, fifth Earl of Carnarvon (1866–1923), who, with the archaeologist Howard Carter, discovered the tomb of Tutankhamun in 1922; Pearl Buck (1892–1973), the American novelist who was the first American woman to win a Nobel prize for literature (in 1938); Laurie Lee (1914–   ), the English poet and novelist; and the Italian conductor Claudio Abbado (1933–   ).

In 1541 Francisco Pizarro, the Spanish conqueror of Peru, was murdered by his former followers.

ᏚᎣ 27 *June*

ON this day ... in 1743 the combined forces of the English, the Hanoverians, and the Austrians, led by George II (the last British king to command an army in the field), defeated the French at the battle of Dettingen; and in 1844 Joseph Smith, the founder of the Mormons, was taken from a gaol in Carthage, Illinois, and shot dead by unknown men.

Born on this day were: Charles Stewart Parnell (1846–91), the Irish patriot and politician known as 'the uncrowned king of Ireland'; Helen Adams Keller (1880–1968), the American writer and scholar who was deaf and blind from the age of nineteen months; the Portuguese cellist Guilhermina Suggia (1888–1950), who was a pupil and mistress of the cellist Casals; and Antoinette Perry (1888–1946), the American actress and producer who gave her name to the 'Tony' awards, presented for outstanding performances in the American theatre.

✌ 28 June

ON this day...in 1914 Gavrilo Princip, a Serbian student, shot and killed Archduke Francis Ferdinand, heir to the throne of Austria, at Sarajevo, starting the chain of events which led to the outbreak of World War I a month later; and in 1919, at Versailles, a peace treaty was signed by Germany and the Allies.

Born on this day were: Henry VIII (1491–1547); Jean Jacques Rousseau (1712–78), the French political philosopher (whose *magnum opus*, *Le Contrat social*, begins with the often-quoted sentence, 'Man is born free, and everywhere he is in chains.'); Luigi Pirandello (1867–1936), the Italian novelist and dramatist and recipient of the 1934 Nobel prize for literature; Alexis Carrel (1873–1944), the French biologist and recipient of the 1912 Nobel prize for physiology and medicine; Pierre Laval (1883–1945), the French premier who was executed for treason after World War II; the American song composer Richard Rodgers (1902–79), especially remembered for the musicals *Oklahoma!*, *The King and I*, *South Pacific*, and *The Sound of Music*; and the English novelist and screenwriter Eric Ambler (1909–   ).

✌ 29 June

THE Flemish painter Peter Paul Rubens (1577–1640) was born on this day, on which the feasts of St Peter and St Paul are celebrated.

In 1888 an English surgeon, Professor Frederick Treves (1853–1923), performed the first appendectomy in Great Britain. (He was later to perform the operation on Edward VII.)

Also born on this day were: William James Mayo (1861–1939), the founder with his brother, Charles Horace Mayo, of the clinic and the foundation for medical education and research in Rochester, Minnesota, named after them; the Italian prima donna Luisa Tetrazzini (1871–1940); Robert Schuman (1886–1963), the French prime minister who proposed and gave his name to the plan for pooling the coal and steel resources of Western Europe; the American singer and film actor Nelson Eddy (1901–67); the American composer Leroy Anderson (1908–75); the American composer Frank Loesser (1910–69); Prince Bernhard of The Netherlands (1911–   ); and the Czech conductor and composer Rafael Kubelik (1914–   ).

✌ 30 June

IN 1934, on this day, the former German chancellor Kurt von Schleicher and Ernst Röhm, the commander of the Brownshirts (the Nazi storm troopers), were among hundreds of leading Nazis murdered in a purge ordered by Adolf Hitler and since known as 'the Night of the Long Knives'. (The assassins were later presented with 'daggers of honour' by the Gestapo chief, Heinrich Himmler, though most of the victims were shot.)

Born on this day were: the English painter Sir Stanley Spencer (1891–1959); the Scottish painter Sir James Gunn (1893–1964); Walter Ulbricht (1893–1973), the East German leader who initiated the building of the Berlin wall in 1961; Harold Laski (1893–1950), the English political scientist and socialist; the English painter Ruskin Spear (1911–90); and the American boxer Mike Tyson (1966–   ), the youngest world heavyweight champion.

In 1690, at the battle of Beachy Head, an Anglo-Dutch fleet commanded by Admiral Torrington was defeated by the French under Admiral Tourville.

In 1859 the Frenchman Charles Blondin made the first crossing of Niagara Falls on a tightrope; and in 1971 three Russian astronauts in *Soyuz II* were found dead on the spacecraft's return to earth. (They were the first men known to have died in space.)

&a 1 July

ON this day ... in 1863 the battle of Gettysburg (the decisive battle of the American Civil War) began; and in 1916 the battle of the Somme began, with more than 21,000 men killed on that first day.

Born on this day were two famous aviators: Louis Blériot (1872–1936), the Frenchman who was the first to fly an aeroplane across the English Channel (in 1909); and Amy Johnson (1903–41), the Englishwoman who was the first airwoman to fly solo from England to Australia (in 1930).

Also born on this day were: Amandine Aurore Lucile Dupin, Baronne Dudevant (1804–76), the French novelist, who took her pseudonym of George Sand from her association with the novelist Jules Sandeau; the British-born film actor Charles Laughton (1899–1962); the British-born film actress Olivia de Havilland (1916–   ); the French-born actress and dancer Leslie Caron (1931–   ); Lady Diana Frances Spencer, HRH The Princess of Wales (1961–   ); and the American Olympic athlete Carl Lewis (1961–   ).

In 1690 James II was defeated by William III at the battle of the Boyne, in Ireland.

&a 2 July

ON this day ... in 1644 Cromwell achieved his first victory over the royalists when he defeated Prince Rupert at Marston Moor, one of the major battles of the Civil War.

In 1881 the US president James Garfield was shot by a disappointed office-seeker, Charles Guiteau, and died the following September. (He was in his first year of office and was the second US president to be assassinated.)

Born on this day were: Thomas Cranmer (1489–1556), archbishop of Canterbury under Henry VIII and Edward VI; the German composer Christoph Willibald von Gluck (1714–87), especially remembered for his opera *Orpheus and Eurydice*; Sir William Henry Bragg (1862–1942), the English scientist who was the joint recipient (with his son Sir William Lawrence Bragg) of the 1915 Nobel prize for physics; and Lord Home of the Hirsel (1903–   ), who as Sir Alec Douglas-Home was the British prime minister 1963–4.

ᵱᴂ 3 *July*

ON this day . . . in 1863, the battle of Gettysburg was effectively ended with the defeat of the Confederate army under General Lee by the Union army under General Meade, with more than 37,000 men killed or wounded over three days.

Born on this day were: Robert Adam (1728–92), the Scottish architect of the eminent Adams family, who gave their name to the neoclassical style of architecture and furniture design; John Singleton Copley (1737–1815), the American portrait painter; the Czech composer Leoš Janáček (1854–1928); the Scottish composer William Wallace (1860–1940); W(illiam) H(enry) Davies (1871–1940), the Welsh poet and 'Super-tramp' (from his *Autobiography of a Super-tramp*), best remembered for his poem beginning, 'What is this life if, full of care I We have no time to stand and stare?'; George M(ichael) Cohan (1878–1942), the American actor, playwright, and songwriter known as 'Yankee Doodle Dandy' (but not 'born on the fourth of July', as claimed in his well-known song); the British playwright and novelist Tom Stoppard (1937–   ); and Sir Richard Hadlee (1951–   ), the New Zealand Test cricketer and the first cricketer to be knighted while still playing Test cricket.

In 1954, food rationing in Britain ended almost nine years after the end of World War II.

ᵱᴂ 4 *July*

TWO US presidents, John Adams and Thomas Jefferson, died on this day in 1826, the fiftieth anniversary of the signing of the American Declaration of Independence; and a third president, James Monroe, died on Independence Day in 1831. Calvin Coolidge (1872–1933) is the only president to have been born on the National Day.

Also born on this day were: Sir George Everest (1790–1866), the surveyor-general of India 1830–43, who gave his name to Mount Everest; Nathaniel Hawthorne (1804–64), the American novelist and short story writer; Giuseppe Garibaldi (1807–82), the Italian patriot and leader of the fight for national independence; Stephen Collins Foster (1826–64), the American songwriter who wrote 'Beautiful Dreamer' and many more highly popular songs; Thomas John Barnardo (1845–1905), the British physician who founded the Barnardo Homes for destitute children; Louis 'Satchmo' (from 'Satchelmouth') Armstrong (1900–71), the American jazz musician and foremost innovator, who learnt to play the cornet in a home for waifs; and Gertrude Lawrence (1902–52), the British actress and singer.

﷽ 5 July

BORN on this day were: Étienne de Silhouette (1709–67), the French minister of finance who gave his name to outline portraits; Luke Hansard (1752–1828), the English printer who gave his name to official reports of parliamentary debates; Cecil John Rhodes (1853–1902), the South African statesman who founded and gave his name to Rhodesia and scholarships at Oxford; and Dwight Filley Davis (1879–1945), the American soldier and politician who donated and gave his name to the international tennis trophy the Davis Cup.

Also born on this day were: Sarah Siddons (1755–1831), the English actress and one of the famous Kemble theatrical family; Sir Thomas Stamford Raffles (1781–1826), the founder of Singapore and London Zoo; Phineas Taylor Barnum (1810–91), the American showman who combined with his rival James Anthony Bailey in founding the famous Barnum and Bailey Circus; and Jean Cocteau (1889–1963), the French poet, playwright, novelist, film director, and influential collaborator in French music.

৯৯ 6 July

ON this day . . . in 1685 the army of James, Duke of Monmouth, claimant to the throne, was defeated by the army of James II at the battle of Sedgemoor, the last major battle to be fought on English soil.

In 1415 the Bohemian religious reformer John Huss was burnt at the stake for heresy. (His execution brought about the subsequent conflicts known as the Hussite War.) In 1535 Henry VIII's former chancellor Sir Thomas More was beheaded for high treason, saying as he mounted the scaffold, 'See me safe up, and for my coming down, let me shift for myself.'

Born on this day were: John Paul Jones (1747–92), the Scottish-born American naval commander who 'invaded' England during the War of Independence; John Flaxman (1755–1826), the English sculptor who designed much of Wedgwood's original pottery; the American actress Nancy Davis (1921–   ), who became President Reagan's second wife; and the Russian-born pianist Vladimir Ashkenazy (1937–   ).

৯৯ 7 July

BORN on this day were: Joseph Marie Jacquard (1752–1834), the Frenchman who invented and gave his name to the first loom to weave patterns successfully and whose use of perforated cards anticipated the modern computer punch-card; the Bohemian-Austrian composer and conductor Gustav Mahler (1860–1911); Marc Chagall (1887–1985), the Russian-born painter whose work is said to have been the first to be described as 'surrealist' (by the French poet Apollinaire, who coined the word); the American film director George Cukor (1899–1983); the Italian film director Vittorio de Sica (1901–74); and the Italian-born American composer Gian-Carlo Menotti (1911–   ).

In 1927 Christopher Stone, the editor of the *Gramophone*, presented the first record programme broadcast by the BBC, thus becoming the first British 'disc jockey' (though he personally disliked the expression).

ஃ 8 July

ON this day . . . in 1709, at one of the decisive battles of the world, Charles XII of Sweden was defeated by Peter the Great's army at Poltava, in the Ukraine, effectively marking the end of the Swedish Empire.

In 1822 the poet Percy Bysshe Shelley was drowned, when his sailing boat sank in a storm in the Bay of Spezia.

Born on this day were: John Davison Rockefeller, senior (1839–1937), founder of the Rockefeller Foundation and other charitable corporations; and his grandson Nelson Aldrich Rockefeller (1908–79), who was the US vice-president 1974–7.

Also born on this day were: the French poet and fabulist Jean de La Fontaine (1621–95); Count Ferdinand von Zeppelin (1838–1917), the German soldier and designer and manufacturer of airships; Sir Arthur John Evans (1851–1941), the English archaeologist who excavated Knossos, the ancient city of Crete; and Percy Grainger (1882–1961), the Australian-born naturalized American composer.

ஃ 9 July

ON this day . . . in 1553 Lady Jane Grey was proclaimed queen of England in succession to Edward VI, who had died three days earlier. (Her 'reign' was ended after thirteen days when she became a prisoner in the Tower of London.)

Born on this day were: Elias Howe (1819–67), the American inventor of the sewing machine; the Yugoslav-born American inventor Nikola Tesla (1856–1943), who gave his name to the unit of magnetic flux density; Ottorino Respighi (1879–1936), the Italian composer; Bruce Bairnsfather (1888–1959), the British artist who created 'Old Bill', the World War I veteran ('. . . if you knows of a better 'ole, go to it.'); Barbara Cartland (1901–   ), the prolific English romantic novelist who is the step-grandmother of the Princess of Wales; the former British prime minister Sir Edward Heath (1916–   ); and the British painter and designer David Hockney (1937–   ).

✧ 10 July

FOUR noted artists were born on this day: the French painter Pierre Joseph Redouté (1759–1840), whose pupils included Marie Antoinette and the Empress Josephine; the French painter Camille Pissarro (1830–1903), one of the leading Impressionists; James Abbott McNeill Whistler (1834–1903), the American portraitist and etcher; and the Italian painter Giorgio de Chirico (1888–1978), whose early work anticipated Surrealism.

Also born on this day were: John Calvin (1509–64), the French theologian and reformer; Frederick Marryat, known as Captain Marryat (1792–1848), the English naval officer and novelist; the French novelist Marcel Proust (1871–1922), whose *magnum opus*, *Remembrance of Things Past*, encompassed thirteen volumes; Edmund Clerihew Bentley (1875–1956), the English journalist and novelist who originated and gave his name to the 'clerihew'; and the American songwriter Jimmy McHugh (1894–1969).

In 1962 the American satellite *Telstar* was launched from Cape Canaveral, Florida, bringing the first live television from the USA to Europe.

✧ 11 July

JOHN Quincy Adams (1767–1848), the sixth US president and son of the second president, John Adams, was born on this day. (His father lived to see him in office and he was the first US president to follow his father into the presidency.)

In 1804 the US vice-president Aaron Burr fought a duel with a former secretary of the treasury, Alexander Hamilton, who was mortally wounded and died the following day.

Also born on this day were: Robert the Bruce (1274–1329), the Scottish king who defeated the army of Edward II at Bannockburn; Thomas Bowdler (1754–1825), the Scottish physician and publisher whose expurgated editions of Shakespeare gave rise to the term 'bowdlerism'; and the English composer and singer Liza Lehmann (1862–1918).

᪥ 12 *July*

ON this day . . . in 1794 Nelson lost his right eye at the siege of Calvi, in Corsica.

In 1910 Charles Stewart Rolls, the English aviation pioneer and co-founder of Rolls-Royce Ltd., was killed in the first fatal aeroplane accident in Britain, little more than a month after he had become the first man to fly the English Channel both ways.

Born on this day were: Josiah Wedgwood (1730–95), founder of the famous pottery works at Etruria, a village in the English Midlands which he built for his workmen and named after the ancient region in Italy; the American writer and philosopher Henry David Thoreau (1817–62), known as 'the hermit of Walden' from the two years he spent living in a hut by Walden Pond, at Concord, Massachusetts; George Eastman (1854–1932), the American inventor of the Kodak camera; the Italian painter Amedeo Modigliani (1884–1920); George Butterworth (1885–1916), the English composer who was killed on the Somme; and Oscar Hammerstein II (1895–1960), the American songwriter and librettist best remembered for his collaboration with Jerome Kern and Richard Rodgers.

᪥ 13 *July*

ON this day . . . in 1793 the French Revolutionary leader Jean Paul Marat was murdered in his bath by a young patriot, Charlotte Corday.

The Austrian composer Arnold Schoenberg died in 1951. (He had been born on 13 September and was obsessed with the fatality of the number to the extent of deliberately misspelling his opera *Moses and Aron.*)

Born on this day were: William Hedley (1779–1843), the English locomotive engineer who invented *Puffing Billy*, the oldest existing steam engine; John Clare (1793–1864), 'the Northamptonshire peasant poet'; Sir George Gilbert Scott (1811–78), the English architect who was the grandfather of the architect Sir Giles Gilbert Scott; Sidney James Webb, Lord Passfield (1859–1947), the English social reformer and economist and a leading member of the Fabian Society; and Eric Williams (1911–83), the

English writer and author of *The Wooden Horse* and other books about POW escapers, who was himself a POW who escaped.

❧ 14 July

ON this day . . . in 1789 the state prison in Paris, the Bastille, was stormed by the populace and the royal troops were forced to surrender their arms. (This action marked the beginning of the French Revolution and has since been known as 'Bastille Day', the French national day.)

Born on this day were: the French statesman Cardinal Mazarin (1602–61), who gave his name to the first printed Bible, since the first known copy was discovered in the library he founded in Paris, the Bibliothèque Mazarine; Emmeline Pankhurst (1858–1928), the English suffragette who led the fight for women's suffrage in Britain by violent means; Captain William Leefe Robinson (1895–1918), the RFC pilot who was awarded the first Victoria Cross to be won in England when he destroyed a Zeppelin in 1916; and the US President Gerald Rudolph Ford (1913–), the first to enter the White House unelected.

In 1881 William H. Bonney, the American outlaw known as 'Billy the Kid', was shot dead by his former friend Patrick Floyd Garrett.

❧ 15 July

ON this day . . . in 1685 the Duke of Monmouth was executed on Tower Hill, nine days after the defeat of his army at Sedgemoor. (It is recorded that his executioner, Jack Ketch, the notorious hangman, took some eight blows to sever Monmouth's head, being relatively inexperienced with the axe.)

In 1857, in the second Cawnpore Massacre, British women and children were murdered (at what is now Kanpur) a few weeks after the massacre of British soldiers and civilians.

Born on this day were: Inigo Jones (1573–1652), the architect known as the 'English Palladio'; the Dutch painter Rembrandt Harmenszoon van Rijn (1606–69); Alfred Harmsworth, Lord Northcliffe (1865–1922), the British newspaper proprietor and a pioneer of mass circulation journalism; the Irish novelist Iris Murdoch (1919–); Julian Bream (1933–), the English guitarist and lutenist; and the English composer Sir Harrison Birtwistle (1934–).

16 July

On this day . . . in 1918 Tsar Nicholas II and his family were murdered by Bolsheviks at Ekaterinburg (now Sverdlovsk).

In 1945 the first atomic bomb was exploded experimentally in New Mexico.

Born on this day were: Sir Joshua Reynolds (1723–92), the English portrait painter and first president of the Royal Academy of Arts; the French landscape painter Jean Baptiste Camille Corot (1796–1875); Mary Morse Eddy (1821–1910), the American founder of the Christian Science Church; Roald Amundsen (1872–1928), the Norwegian explorer who led the first successful expedition to the South Pole; Bela Schick (1877–1967), the Hungarian paediatrician who discovered and gave his name to the test for determining susceptibility to diphtheria; the American film actress Barbara Stanwyck (1907–90); the American film actress and dancer Ginger Rogers (1911–); and the Israeli-born violinist Pinchas Zuckerman (1948–).

17 July

On this day . . . in 1453 the Hundred Years War between England and France was ended with the defeat of the English at the battle of Castillon, leaving Henry VI with Calais alone of all his former French possessions.

In 1762 Peter III, the Russian emperor, was murdered (probably with the connivance of his wife, who succeeded him as Catherine II).

In 1793 Charlotte Corday, the assassin of the French Revolutionary leader Jean Paul Marat, was sent to the guillotine.

Born on this day were: Isaac Watts (1674–1748), the English hymnwriter; Sir Donald Tovey (1875–1940), the English composer and writer on music; Erle Stanley Gardner (1889–1970), the American lawyer and writer of detective fiction who created the lawyer 'Perry Mason'; and the American film actor and dancer James Cagney (1899–1986).

❧ 18 July

ON this day . . . in 1936 the Spanish Civil War began as General Franco led the Fascist revolt of army chiefs against the Republican government.

Born on this day were: Gilbert White (1720–93), the English clergyman and naturalist who immortalized the village of Selborne, in Hampshire, in his writing; the English novelist William Makepeace Thackeray (1811–63); William Gilbert Grace (1848–1915), the England cricketer known simply as 'W.G.' or 'The Great Cricketer', as he is described on the gates named after him at Lord's Cricket Ground; Hendrick Antoon Lorentz (1853–1928), the Dutch scientist who shared the 1902 Nobel prize for physics with Pieter Zeeman; the English writer Laurence Housman (1865–1959), the younger brother of A. E. Housman; Vidkun Quisling (1887–1945), the Norwegian politician whose name became synonymous with traitorousness from his collaboration with the Nazis in World War II; Nelson Mandela (1918–   ), the African nationalist leader who was sentenced to life imprisonment in 1964 and not released until 1990; John Glenn (1921–   ), the American astronaut who was the first to orbit the earth (in 1962); and the Russian poet and dissident Yevegeny Aleksandrovich Yevtushenko (1933–   ).

❧ 19 July

ON this day . . . in 1545 Henry VIII watched his flagship *Mary Rose* capsize and sink in the Solent as it prepared to go into battle against the French; and in 1553 Henry's daughter Mary became queen of England with the imprisonment of Lady Jane Grey at the end of her 'reign' of thirteen days.

Born on this day were: Samuel Colt (1814–62), the American inventor of the first practical revolver; the French painter Edgar Degas (1834–1917); Charles Horace Mayo (1865–1939), the American surgeon who with his brother was the co-founder of the Mayo Foundation for Medical Education and Research; the Scottish novelist and doctor of medicine A(rchibald) J(oseph) Cronin (1896–1981), who created 'Dr Finlay'; the English songwriter and actor Hubert Gregg (1916–); and the English painter and writer John Bratby (1928–92).

৯৯ 20 July

ON this day . . . in 1944 Adolf Hitler survived an attempt on his life organized by senior German officers and others.

In 1969 the American astronauts Neil Armstrong and Edwin 'Buzz' Aldrin landed on the moon in the lunar module *Eagle*, at 9.18 p.m. BST.

Born on this day were: the Italian poet Francesco Petrarch (1304–74); John Sholto Douglas, eighth Marquis of Queensberry (1844–1900), the patron of boxing who gave his name to the 'Queensberry rules'; Alberto Santos-Dumont (1873–1932), the Brazilian aviation pioneer; John Reith, Lord Reith (1889–1971), the British statesman who was the first director-general of the BBC; and Sir Edmund Hillary (1919–), the New Zealand mountaineer who with the Sherpa Tenzing Norgay became the first to reach the summit of Mount Everest, in 1953.

In 1951 King Abdullah of Jordan was assassinated.

৯৯ 21 July

ON this day . . . in 1403 the Percys were defeated by Henry IV and Sir Henry Percy, known as 'Hotspur', killed at the battle of Shrewsbury; Napoleon defeated the Mamelukes in Egypt at the battle of the Pyramids in 1796; and in 1861 the Union army was defeated by the Confederates at the battle of Bull Run, in Virginia, and Brigadier-General Thomas Jonathan Jackson received his famous nickname from a fellow officer. ('See, there is Jackson standing like a stonewall.')

In 1960 Mrs Sirimavo Bandaranaike became the first woman prime minister as prime minister of Ceylon.

Born on this day were: Thomas Pelham-Holles, first Duke of Newcastle (1693–1768), the British prime minister who succeeded his brother in office in 1754; Israel Beer, later Baron Paul Julius von Reuter (1816–99), the German-born naturalized British pioneer of the news agency; and Sir C(harles) Aubrey Smith (1863–1948), the England cricket captain who became internationally famous as a film actor.

In 1969, at 02.56 GMT, the American astronaut Neil Armstrong became the first man to walk on the moon, saying to the millions of television viewers on earth, 'That's one small step for man, one giant leap for mankind.'

৪৯ 22 July

ON this day . . . in 1910 wireless telegraphy was first used to assist the apprehension of a criminal when the captain of SS *Montrose*, bound for Canada, sent a message back to England concerning two of his passengers, a Mr Robinson and his 'son', who were in fact the suspected murderer Hawley Harvey Crippen and his mistress Ethel Le Neve.

In 1934 the American gangster John Herbert Dillinger was shot dead by FBI officers outside a Chicago cinema.

Born on this day were: Thomas Stevenson (1818–87), a son of the Scottish engineer Robert Stevenson and father of the writer Robert Louis Stevenson, who, like his father and his brothers Alan and David, 'made the great sea lights in every quarter of the world . . . shine more brightly'; Alfred Percival Graves (1846–1931), the Irish poet and songwriter and father of the writer Robert Graves; Gwen John (1876–1939), the British painter and sister of the painter Augustus John.

Also born on this day were: William Archibald Spooner (1844–1930), the English clergyman who gave us Spoonerisms; and Gregor Johann Mendel (1822–84), the Austrian botanist who gave us a theory of heredity, Mendelism.

৪৯ 23 July

ON this day . . . in 1865 'General' William Booth founded the religious movement which, thirteen years later, became known as the Salvation Army.

Born on this day were: the English poet Coventry Patmore (1823–96); Sir Arthur Whitten-Brown (1886–1948), the British airman who in 1919 made the first non-stop transatlantic flight with (Sir) John Alcock; Raymond Chandler (1888–1959), the American writer of detective fiction and creator of 'Philip Marlowe'; Ras Tafari, Emperor Haile Selassie of Ethiopia (1891–1975), whose name has become identified with Rastafarianism; and Michael Foot (1913–   ), the leader of the British Labour Party, 1980–3.

❧ 24 July

ON this day . . . in 1704 Admiral Sir George Rooke took Gibraltar from the Spaniards and gained a new possession for Britain.

In 1797 Nelson received the wound that led to the loss of his right arm in an unsuccessful assault on Santa Cruz de Tenerife.

Amelia Earhart (1898–1937), the American aviatrix who was the first woman to cross the Atlantic in an aeroplane, was born on this day; and in 1883 Matthew Webb (known as Captain Webb), the first man to swim the English Channel, was drowned in attempting to swim the rapids below Niagara Falls.

Also born on this day were: Simón Bolívar (1783–1830), the founder of Bolivia; Alexandre Dumas (1802–70), the French novelist and playwright especially remembered for *The Three Musketeers*; and Robert Graves (1895–1985), the English poet and novelist who in his autobiographical *Goodbye to All That* relates how he read his 'obituary' in *The Times* on his 21st birthday, after being wounded in World War I.

❧ 25 July

ON this day . . . in 1909 the French aviator Louis Blériot became the first man to fly the English Channel in a heavier-than-air machine, when he crossed from Calais to Lydd, in Kent, in an aeroplane of his own design.

Born on this day were: Arthur James Balfour (1848–1930), later first Earl of Balfour, the British prime minister (1902–5) who, as Foreign Secretary, in 1917, made the declaration (since named after him) that the British government favoured the establishment of a Jewish national home in Palestine; and Gavrilo Princip (1894–1918), the Serbian student who assassinated Archduke Ferdinand and his wife at Sarajevo in 1914 (the incident leading to the outbreak of World War I).

In 1934 the Austrian chancellor Engelbert Dollfuss was shot and killed by Nazis, who seized the chancellery.

In 1978 the first 'test-tube' baby, Louise Brown, was born in a hospital in Oldham, in England.

✌ 26 July

On this day . . . in 1847 Liberia became the first African colony to become an independent state, and 26 July is the republic's national day.

Born on this day were: John Field (1782–1837), the Irish composer who invented the nocturne; the Irish dramatist George Bernard Shaw (1856–1950), who was also a musician and a noted music critic as 'Corno di Bassetto'; Serge Koussevitsky (1874–1951), the Russian-born American conductor, composer, and double-bass virtuoso; Carl Jung (1875–1961), the Swiss psychoanalyst and former disciple of Freud; the French writer André Maurois (1885–1967); the English novelist Aldous Huxley (1894–1963); the American film writer, producer, and director Blake Edwards (1922–   ); and the American film writer, producer, and director Stanley Kubrick (1928–   ).

In 1918 Major Edward 'Mick' Mannock, VC, the British airman credited with shooting down the most German aircraft in World War I, was shot down and killed by fire from the ground.

✌ 27 July

On this day . . . in 1689 John Graham of Claverhouse (known variously as 'Bonnie Dundee' and 'Bloody Claverse') was mortally wounded at the battle of Killiecrankie as he led the Jacobites against King William's army.

In 1778 the first battle of Ushant took place between the British and French fleets with no clear victory for either side and a court martial for the British Admiral Keppel (who was eventually acquitted and became First Lord of the Admiralty).

Born on this day were: Alexandre Dumas, 'fils' (1824–95), the French novelist and playwright, best known for *La Dame aux camélias*; the Spanish composer Enrique Granados (1867–1916); Hilaire Belloc (1870–1953), the French-born English writer of great versatility; the Hungarian composer Ernö von Dohnányi (1877–1960); Sir Geoffrey de Havilland (1882–1965), the British aircraft designer and manufacturer who gave his name to many famous civil and military aircraft spanning both World Wars; and Sir Anton Dolin (1904–83), the British dancer and choreographer.

In 1953 the Korean War ended with the signing of an armistice at Panmunjon.

ఊ 28 *July*

On this day ... in 1914 World War I began with Austria-Hungary declaring war on Serbia, exactly one month after the assassination of Archduke Ferdinand at Sarajevo.

Born on this day were: Sir Hudson Lowe (1769–1844), the British soldier who was governor of St Helena 1815–21 and custodian of Napoleon; Gerard Manley Hopkins (1844–89), the English poet and Jesuit priest, none of whose poems were published in his lifetime; Beatrix Potter (1866–1943), the author and illustrator of children's books; Sir Karl Popper (1902–   ), the Austrian-born philosopher; Jacqueline Onassis (1929–   ), widow of US president John F. Kennedy and shipowner Aristotle Onassis; and Sir Garfield St Aubrun 'Gary' Sobers (1936–   ), the former West Indies cricket captain and one of the greatest all-round cricketers.

In 1540 Henry VIII married his fifth wife Catherine Howard, on the same day that his former chief minister, Thomas Cromwell, was beheaded on Tower Hill; and in 1794 Robespierre, Saint-Just, and nineteen other French Revolutionaries were sent to the guillotine.

ᢀ 29 July

ON this day . . . in 1588 the Spanish Armada was first sighted off the coast of Cornwall; while according to the Old Style calendar, still in use in England at that time, the Armada was finally routed. ('God blew and they were scattered.')

Born on this day were: George Bradshaw (1801–53), the English printer who originated and gave his name to the first national railway timetable; the Norwegian physician Armauer Gerhard Henrik Hansen (1841–1912), discoverer of the bacillus causing leprosy, Hansen's bacillus; Booth Tarkington (1869–1946), the American novelist and playwright; the Italian dictator Benito Mussolini (1883–1945); Sigmund Romberg (1887–1951), the Hungarian-born composer best remembered for his operettas *The Student Prince* and *The Desert Song*; and the Greek composer Mikis Theodorakis (1925–   ), best remembered for his music for the film *Zorba the Greek*.

The German composer Robert Schumann died in an asylum, aged 46, after attempting suicide; and the Dutch painter Vincent van Gogh died in 1890, aged 37, two days after shooting himself.

ᢀ 30 July

EMILY Jane Brontë (1818–48), the English novelist and poet especially remembered for her novel *Wuthering Heights*, was born on this day; and the English poet Thomas Gray died in 1771 and was buried (beside his mother) at Stoke Poges, the village in Buckinghamshire which inspired his 'Elegy Written in a Country Churchyard'. ('No farther seek his merits to disclose, I Or draw his frailties from their dread abode . . .')

Also born on this day were: Henry Ford (1863–1947), the American motor car manufacturer and pioneer of the assembly line production method; Henry Moore (1898–1986), the English sculptor; Gerald Moore (1899–1987), the English pianist and leading accompanist of his time; Cyril Northcote Parkinson (1909–   ), the English historian and political economist who propounded and gave his name to Parkinson's law ('Work expands so as to fill the time available for its completion.'); and the British Olympic decathlon gold medallist Daley Thompson (1958–   ).

ॐ 31 July

ON this day . . . in 1917 the third battle of Ypres began and fourteen Victoria Crosses were won, including the second VC to be won by Captain Godfrey Chavasse, who died of his wounds; and in World War II, in 1943, Captain Hedley Verity, the Yorkshire and England cricketer, died of wounds in Italy.

Born on this day were: John Ericsson (1803–89), the Swedish inventor who pioneered the introduction of screw propellers and designed armoured warships for the US Navy; George Henry Thomas (1816–70), the American Unionist general known as 'the Rock of Chickamauga' from his heroic stand at the battle on the Georgia–Tennessee border in 1863; Milton Friedman (1912–   ), the American economist and recipient of the 1976 Nobel prize for economics and guru of the 'monetarists'; and the English conductor and musicologist Norman Del Mar (1919–   ).

AUGUST ✑

✑ 1 August

On this day . . . in 1740 Thomas Arne's masque *Alfred*, including the patriotic song 'Rule, Britannia!', was first performed at the country house of the Prince of Wales, in Buckinghamshire.

In 1798 Nelson routed the French fleet at the battle of the Nile, in Aboukir Bay.

Born on this day were two American writers especially associated with seafaring: Richard Henry Dana (1815–82), whose *Two Years before the Mast* was an account of his personal experiences as a common sailor; and Herman Melville (1819–91), the author of the novels *Moby Dick* and *Billy Budd*, who was variously a whaler and a seaman on a man-of-war.

Also born on this day were: Jean Baptiste Pierre Antoine de Monet Lamark (1744–1829), the French naturalist who originated and gave his name to a theory of evolution which anticipated Darwinism; Francis Scott Key (1779–1843), the American lawyer who wrote 'The Star Spangled Banner', the official national anthem of the USA since 1931; and the British songwriter Lionel Bart (1930–   ), who wrote the musical play *Oliver!*

In 1914 Germany declared war on Russia, three days before Britain declared war on Germany.

✑ 2 August

On this day . . . in 1100 William II (called 'Rufus') was killed by an arrow while hunting in the New Forest and was succeeded by his brother Henry I.

In 1934 the German president Hindenburg died and Hitler assumed the title of 'Der Führer'.

Born on this day were: John Tyndall (1820–93), the British scientist who discovered (among other things) why the sky is blue; Elisha Gray (1835–1901), the American inventor who claimed that his invention of a telephone preceded that of Alexander Graham Bell; the English poet Ernest Dowson (1867–1900), whose two best-remembered poems have inspired the titles of a famous novel (*Gone with the Wind*) and a memorable song from a film of the same name ('The Days of Wine and Roses'); the English novelist Ethel M(ary) Dell (1881–1939); and Sir Arthur Bliss (1891–1975), the English composer who was Master of the Queen's Music from 1953.

ℰ 3 August

ON this day . . . in 1914 Germany declared war on France (the day before Britain declared war on Germany).

In 1916 the former British consular official Sir Roger Casement was hanged for treason, at Pentonville prison.

Born on this day were: Stanley Baldwin (1867–1947), three times British prime minister and first Earl Baldwin; Rupert Brooke (1887–1915), the British 'war poet' who died of blood poisoning in Greece on his way to the Dardanelles; P(hyllis) D(orothy) James, Baroness James (1920–), the English writer of crime fiction; and the American singer Tony Bennett (1926–).

In 1492 Christopher Columbus set sail in the *Santa Maria* (with the *Pinta* and the *Niña*) on the first of his four famous voyages; and in 1958 the American submarine *Nautilus*, the first nuclear submarine, passed under the North Pole.

ℰ 4 August

ON this day . . . in 1265 Simon de Montfort, Earl of Leicester and brother-in-law of Henry III (and virtually the founder of the English Parliament), was defeated and killed at the battle of Evesham by Henry's son Prince Edward, later Edward I.

In 1914 Germany invaded Belgium and Britain declared war on Germany. ('The lamps are going out all over Europe; we shall not see them lit again in our lifetime'—Sir Edward Grey.)

Born on this day were: Percy Bysshe Shelley (1792–1822), the English poet; Edward Irving (1792–1834), the Scottish clergyman who founded the Holy Catholic Apostolic Church (the 'Irvingites'); Walter Pater (1839–94), the English writer and critic; William Henry Hudson (1841–1922), the British writer on natural history who is commemorated by a bird sanctuary in Hyde Park; Sir Harry Lauder (1870–1950), the Scottish comedian and songwriter; Dame Laura Knight (1877–1970), the English artist; and HM Queen Elizabeth, The Queen Mother (1900–   ), the mother of Queen Elizabeth II.

✇ 5 *August*

ON this day . . . in 1914 the day after Britain declared war on Germany, the mobilization of the BEF (the British Expeditionary Force) began. (Its nickname, 'the Old Contemptibles', came from Kaiser Wilhelm's reference to 'General French's contemptible little army'.)

In 1916 the English composer George Butterworth was killed in action at the battle of the Somme.

Born on this day were: Guy de Maupassant (1850–93), the French novelist and short-story writer; Louis Wain (1860–1939), the English artist and caricaturist (especially of cats); and Neil A. Armstrong (1930–   ), the American astronaut who was the first man to walk on the moon (see 21 July).

The American film actress Marilyn Monroe died in 1962, aged 36.

✇ 6 *August*

ON this day . . . in 1945 the first atomic bomb was dropped on Hiroshima.

Born on this day were: Daniel O'Connell (1775–1847), the Irish political leader known as 'the Liberator'; Dorothy ('Dora') Wordsworth (1804–47), the author and daughter of the Poet Laureate William Wordsworth; the Poet Laureate Alfred, Lord Tennyson (1809–92), who succeeded Wordsworth as Laureate in 1850; James Henry Greathead (1844–96), the South African-born engineer who constructed the first tube railway in London; Paul Claudel (1868–1955), the French poet and diplomat; Frederick Thomas Jane (1870–1916), the British journalist and artist who founded and gave his name to *Jane's Fighting Ships* and *All the World's Aircraft*; Sir Alexander Fleming (1881–1955), the Scottish bacteriologist who in 1928 dicovered penicillin; the American film actor Robert Mitchum (1917–   ); and Chris Bonington (1934–   ), the British mountaineer and writer.

In 1888 a 35-year-old prostitute, Martha Turner, believed to have been the first victim of 'Jack the Ripper', was stabbed to death in Whitechapel, in London's East End.

✌ 7 *August*

On this day . . . in 1913 'Colonel' Samuel Franklin Cody, the American-born airman who made the first 'official' flight in an aeroplane in England (in 1908), was killed in a flying accident.

Born on this day were: Frederick William Farrar (1831–1903), a dean of Canterbury and author of the popular novel *Eric, or Little by Little*, who was the grandfather of Lord Montgomery of Alamein; the English composer Sir Granville Bantock (1868–1946); Margaretha Geertruida Zelle (1876–1917), the Dutch dancer and alleged spy known as Mata Hari; Dornford Yates (1885–1960), the English novelist whose real name was Cecil William Mercer; Louis Leakey (1903–72), the British anthropologist whose discoveries put man's ancestry back by some 20 million years; and Dr Ralph Bunche (1904–71), the UN mediator who was the recipient of the 1950 Nobel peace prize and the first black man to receive this award.

In 1931 Bix Beiderbecke, the legendary jazz cornet player, died at the age of 28.

✌ 8 *August*

On this day . . . in 1940 the first of a series of intensive air raids on Great Britain, which were to continue into the following October, marked the beginning of the battle of Britain.

In 1945 Russia declared war on Japan.

Born on this day were: Gottfried Kniller, later Sir Godfrey Kneller (1646–1723), the German-born portrait painter who gave his name to Kneller Hall, his London residence, which became the home of the Royal Military School of Music; Thomas Anstey Guthrie (1856–1934), the English novelist and playwright who used the pseudonym F. Anstey; Charles (Harold St John) Hamilton (1876–1961), the English writer known as Frank Richards (one of his many pseudonyms), who created 'Billy Bunter', the fat schoolboy of the comic papers; Ernest Orlando Lawrence (1901–58), the American atomic physicist and recipient of the 1939 Nobel prize for physics; and the American actor Dustin Hoffman (1937–   ).

৯৯ 9 August

ON this day ... in 480 BC the battle of Thermopylae came to its end when Leonidas, king of Sparta, and some 300 men were finally overcome by a vast Persian army, after a resistance of three days.

In 1945, three days after the first atomic bomb was dropped on Hiroshima, a second was dropped on Nagasaki.

Born on this day were: Henry V (1387–1422), the English king who led his army in the battle of Agincourt; Izaak Walton (1593–1683), the English writer and biographer especially remembered for *The Compleat Angler, or the Contemplative Man's Recreation*; John Dryden (1631–1700), the first Poet Laureate to hold the title officially; Thomas Telford (1757–1834), the Scottish engineer who built bridges, canals, and roads; Albert Ketèlbey (1875–1959), the English composer who wrote *In a Monastery Garden* and *In a Persian Market*; the English pianist Solomon (Cutner), whose career was ended by a stroke in 1965 (1902–88); and Philip Larkin (1922–86), the English poet.

In 1974 Richard Nixon became the first US president to resign the office.

৯৯ 10 *August*

ON this day . . . in 1628 the Swedish warship *Vasa* capsized and sank in Stockholm harbour at the start of her maiden voyage. (Some 300 years later the ship was recovered from the seabed and restored.)

Born on this day were: Camillo Benso di Cavour (1810–61), the Italian statesman and unifier of Italy; Charles Samuel Keene (1823–91), the English illustrator who contributed to *Punch* for forty years; William Willett (1856–1915), the English pioneer of 'daylight saving', who died the year before BST (British Summer Time) was adopted in Great Britain; the Russian composer Alexander Konstantinovich Glazunov (1865–1936); the English poet Laurence Binyon (1869–1943), especially remembered for his poem to the dead of World War I ('They shall grow not old, as we that are left grow old . . .'); and Herbert Clark Hoover (1874–1964), the thirty-first US president.

In 1895 Henry Wood (he was not knighted until 1911) conducted the first Promenade Concert at the old Queen's Hall.

৯৯ 11 *August*

ON this day . . . in 1711 the first race meeting at Ascot was held in honour of Queen Anne.

The Scottish-born multimillionaire and philanthropist Andrew Carnegie died in 1919, having given more than £70,000,000 in benefactions, including the provision of libraries in Great Britain and the United States. ('A man who dies rich dies disgraced.')

Born on this day were: the English novelist Charlotte Mary Yonge (1823–1901); Bertram Mills (1873–1938), the British circus proprietor whose shows were the first to be televised, in 1938, the year of his death; Enid Blyton (1897–1968), the writer of stories for children who was the third most translated British author after Agatha Christie and Shakespeare; Sir Angus Wilson (1913–91), the British novelist and critic; and the Welsh composer Alun Hoddinott (1929–).

✤ 12 *August*

ON this day . . . in 1822 Lord Castlereagh, the British Foreign Secretary, committed suicide by cutting his throat with a penknife.

Born on this day were: Thomas Bewick (1753–1828), the English woodengraver and naturalist; George IV (1762–1830), who became Prince Regent in 1810 in consequence of the insanity of his father George III; Robert Southey (1774–1843), the Poet Laureate from 1813; John Francis Stanley Russell, the second Earl Russell (1865–1931), older brother of Bertrand Russell and the first peer to become a member of the Labour Party (and the first holder of a British car registration, 'A1'); Cecil B(lount) De Mille (1881–1959), the American film producer; the English novelist Frank Swinnerton (1884–1982); Cyril Joad (1891–1953), the philosopher and popular broadcaster who coined the phrase 'It all depends on what you mean by . . .'; Wing Commander Guy Gibson, VC (1918–1944), the British airman who led the 'Dam Busters' raid in 1943; and Norris McWhirter (1925–), the editor and compiler of *The Guinness Book of Records* (with his twin brother the late Ross McWhirter).

In 1961 the erection of the Berlin Wall began, surrounding the Western sector of the city.

✤ 13 *August*

ON this day . . . in 1704 the Duke of Marlborough, with Prince Eugene of Austria, defeated the French under Marshal Tallard at the battle of Blenheim. ('And everybody praised the Duke | Who this great fight did win.' | 'But what good came of it at last?' | Quoth little Peterkin:— | 'Why, that I cannot tell,' said he, | 'But 'twas a famous victory.'—Robert Southey.)

Born on this day were: Queen Adelaide (1792–1849), the consort of William IV, who gave her name to the capital of South Australia; Sir George Grove (1820–1900), the English engineer who became the first director of the Royal College of Music and the first editor of (Grove's) *Dictionary of Music and Musicians*; Phoebe Anne Moses (1860–1926), known as Annie Oakley, the American markswoman and entertainer and eponymous heroine of Irving Berlin's *Annie Get Your Gun*; the

English composer John Ireland (1879–1962); Sir Alfred Hitchcock (1899–1980), the English film director; Sir Basil Spence (1907–76), the Scottish architect who designed the new Coventry Cathedral; Ben Hogan (1912–), the American golfer who won every major world championship; George Shearing (1919–), the English-born jazz pianist and composer; and Fidel Castro (1927–), the president of the Republic of Cuba since 1976.

ᔐ 14 *August*

On this day . . . in 1945 Japan surrendered unconditionally to the Allies.

In 1947 Pakistan became an independent dominion.

Born on this day were: Hans Christian Oersted (1777–1851), the Danish scientist who discovered electromagnetism and gave his name to the unit of magnetic field strength; Samuel Sebastian Wesley (1810–76), the English organist and composer and natural son of the composer Samuel Wesley; the English novelist and playwright John Galsworthy (1867–1933), who won the 1932 Nobel prize for literature in the last year of his life; and the English novelist and playwright Frederic Raphael (1931–).

ᔐ 15 *August*

On this day . . . in 1872 the first voting by ballot in Great Britain took place in a by-election at Pontefract, when Hugh Childers, a Liberal MP and a minister, was re-elected.

In 1945 the surrender of Japan was celebrated as VJ (Victory over Japan) Day.

Born on this day were: Napoleon Bonaparte (1769–1821); Sir Walter Scott (1771–1832), the Scottish novelist and poet; Thomas De Quincey (1785–1859), the English writer especially remembered for his *Confessions of an English Opium Eater*; James Keir Hardie (1856–1915), the Scottish Labour leader and the first British Labour candidate; Samuel Coleridge-Taylor (1875–1912), the English composer of African descent; Edna Ferber (1887–1968), the American writer whose novel *Show Boat* inspired the Kern/Hammerstein musical play of the same name; T(homas) E(dward) Lawrence (1888–1935), the British soldier and Arabist who became known as 'Lawrence of Arabia' in World War I; and Princess Anne, the Princess Royal (1950–   ).

In 1935 Will Rogers, the American actor and humorous writer known as 'the cowboy philosopher', was killed in an aeroplane crash with Wiley Post, the American aviation pioneer.

✥ 16 *August*

ON this day . . . in 1513 Henry VIII commanded his troops at Guinegatte in a victory over the French, since known as the battle of the Spurs; and in 1819 a reform meeting at St Peter's Field, in Manchester, was brought to a violent end with a charge of the yeomanry, including some Waterloo veterans. (The incident was consequently named 'the Peterloo Massacre'.)

Born on this day were: Jean de La Bruyère (1645–96), the French writer remembered for his maxims, such as, 'There are some who speak one moment before they think.'; Frederick Augustus, Duke of York (1763–1827), second son of George III, who was parodied in the rhyme 'The Grand Old Duke of York'; Lady Nairne (1766–1845), the Scottish poet and author of many Jacobite songs, such as 'Will ye no come back again' and 'Charlie is my Darling'; the English novelist Georgette Heyer (1902–74); and Menachem Begin (1913–92), the Israeli statesman who with President Sadat of Egypt was awarded the 1978 Nobel peace prize. The American popular singer Elvis Presley, known as 'the king of rock 'n' roll', died in 1977, aged 42.

✥ 17 *August*

ON this day . . . in 1812 Napoleon's army defeated the Russians at the battle of Smolensk, on the retreat to Moscow.

In 1833 the first steamship to cross the Atlantic entirely under power, the Canadian vessel the *Royal William*, began her journey from Nova

Scotia to the Isle of Wight; and in 1978 three Americans completed the first crossing of the Atlantic in a balloon.

Born on this day were: David (known as 'Davy') Crockett (1786–1836), the American frontiersman and politician who gave his name to a hat; Mae West (1892–1980), the American actress and sex symbol who gave her name to a pneumatic life-jacket; and Edward James 'Ted' Hughes (1930–   ), the English poet who was married to the American poet Sylvia Plath and who succeeded Sir John Betjeman as Poet Laureate in 1984.

ᕙ 18 *August*

ON this day . . . in 1759 the French fleet was defeated by the British under Admiral ('Old Dreadnought') Boscawen at the battle of Lagos Bay; and at the battle of Gravelotte (one of the major battles of the Franco-Prussian War) the French under Marshal Bazaine were forced to retreat to Metz.

Born on this day were: Antonio Salieri (1750–1825), the Italian composer who is alleged to have been involved in the death of Mozart (though there is no real evidence of this); Franz Joseph I (1830–1916), the Austrian emperor whose attack on Serbia in 1914 precipitated World War I; Basil Cameron (1884–1975), the English conductor who in 1912 changed his name to Hindenburg (but understandably reverted to his original name at the outbreak of the War); the English concert pianist Moura Lympany (1916–   ); the American film actress Shelley Winters (1922–   ); and the American film actor Robert Redford (1937–   ).

ᕙ 19 *August*

ON this day . . . in 1942 a large-scale raid on the port of Dieppe by British and Canadian troops resulted in high casualties for the attacking force. (The town was eventually liberated in 1944 by Canadian soldiers who had survived the 1942 raid.)

In 1936 the Spanish poet Federico García Lorca was shot by Franco's sympathizers, after being forced to dig his own grave.

Born on this day were: John Flamsteed (1646–1719), the first Astronomer Royal (in 1675), for whom Charles II built the observatory at Greenwich; James Nasmyth (1808–90), the Scottish engineer who invented the steam hammer; Edith Nesbit (1858–1924), the English novelist and poet and writer of books for children; Orville Wright (1871–1948), the American pioneer of aviation who (with his brother Wilbur) made the first powered flight to be officially recognized (on 17 December 1903); George Enescu (1881–1955), the Romanian violinist and composer and mentor of Sir Yehudi Menuhin; the American humorous writer Ogden Nash (1902–71); and Bill Clinton (1945–   ), the forty-second US president.

✌ 20 August

ON this day . . . in 1940 Leon Trotsky, the exiled Russian revolutionary, was attacked by a hired assassin in Mexico City and died the next day.

Rajiv Gandhi (1944–91), the prime minister of India who was assassinated less than seven years after the assassination of his mother Indira Gandhi, was born on this day.

Also born on this day were: Jacopo Peri (1561–1633), the Italian composer generally considered to have composed the first known operas; Benjamin Harrison (1833–1901), the twenty-third US president and grandson of the ninth president, William Henry Harrison; Raymond Poincaré (1860–1934), the French statesman and ninth president of the Republic who was also three times prime minister of France; and Jack Teagarden (1905–64), the American jazz trombone player and singer and one of the pioneers of jazz music.

✌ 21 August

ON this day . . . in 1808 Napoleon's General Junot was defeated by Wellington (then Arthur Wellesley) at the battle of Vimiero, the first battle of the Peninsular War.

Born on this day were: William IV (1765–1837), the third son of George III; and HRH The Princess Margaret, Countess of Snowdon (1930–), the second daughter of George VI and Queen Elizabeth (known in her youth as Princess Margaret Rose).

Also born on this day were: William Murdoch (1754–1839), the Scottish engineer who invented coal-gas lighting; the artist and illustrator Aubrey Beardsley (1872–98), one of the leaders of the 'Decadent' movement of the 1890s; Claude Grahame-White (1879–1959), the pioneer aviator and the first Englishman to receive a certificate of proficiency in aviation; and another famous airman, Captain Albert Ball, VC (1896–1917), who when he died (in combat) had won more honours for bravery than anyone of his age had previously achieved.

৪১ 22 August

ON this day . . . in 1485 Richard III's army was defeated and Richard was killed at Bosworth Field, in a battle with the army of Henry, Earl of Richmond, who (on this same day) began his reign as Henry VII. (This battle also brought an end to the Wars of the Roses.)

In 1642 the Civil War in England began as Charles I raised his standard at Nottingham.

Born on this day were: the Austrian composer and conductor Joseph Strauss (1827–70), the second son of Johann Strauss the elder; the Scottish composer Sir Alexander Mackenzie (1847–1935); the French composer Claude Debussy (1862–1918); and the German composer and proponent of electronic music Karlheinz Stockhausen (1928–).

Also born on this day was the American satirical writer Dorothy Parker (1893–1967), who said (among many other things), 'Why is it no one ever sent me yet I One perfect limousine, do you suppose? I Ah no, it's always just my luck to get I One perfect rose.'

℀ 23 August

ON this day... in 1305 the Scottish patriot Sir William Wallace was hanged, drawn, beheaded, and quartered in London; the quarters then being sent to Newcastle, Berwick, Stirling, and Perth respectively.

Louis XVI (1754–93), the French king who was to die by the guillotine, was born on this day.

Also born on this day were: W(illiam) E(rnest) Henley (1849–1903), the English poet and critic; Arnold Toynbee (1852–83), the English historian and social reformer who gave his name to the world's first social settlement and coined the phrase 'the industrial revolution'; Constant Lambert (1905–51), the English composer, conductor, and musicologist; the early music specialist Carl Dolmetsch (1911–   ); and Gene Kelly (1912–   ), the American dancer, singer, and film actor.

In 1914, as Japan declared war on Germany, the British Expeditionary Force fought its first battle at Mons (and the first Victoria Crosses of World War I were earned).

℀ 24 August

ON this day... in AD 79 the Italian city of Pompeii was completely buried by the eruption of Mount Vesuvius.

In 1572 some 50,000 people were put to death in 'the Massacre of St Bartholomew', when Charles IX, under the influence of his mother Catherine de Medici, ordered the massacre of the Huguenots throughout France on the day of the feast of St Bartholomew.

Born on this day were: George Stubbs (1724–1806), the English painter who wrote and illustrated *The Anatomy of the Horse*; William Wilberforce (1759–1833), the English philanthropist and slave-trade abolitionist; Sir Max Beerbohm (1872–1956), the English writer and caricaturist; the English painter Graham Sutherland (1903–80); and the English poet Charles Causley (1917–   ).

❧ 25 August

ON this day . . . in 1875 the English seaman known as 'Captain' Matthew Webb became the first man to swim across the English Channel (from Dover to Calais, in 21¾ hours).

In 1770 Thomas Chatterton, the English poet and literary forger (immortalized by Wordsworth as 'Chatterton, the marvellous boy, | The sleepless soul, that perished in his pride'), died by his own hand at the age of 17.

Born on this day were: Ivan IV (1530–84), the Russian ruler (and the first to assume the title of 'tsar') who was known as 'Ivan the Terrible' from the atrocities he inflicted on his subjects (including the killing of his eldest son in a fit of rage); Allan Pinkerton (1819–84), the Scotsman who founded the first private detective agency in America; Robert Stolz (1880–1975), the Austrian composer and conductor best remembered for his operetta *The White Horse Inn*; and the American composer and conductor Leonard Bernstein (1918–90), best remembered for *West Side Story*.

❧ 26 August

ON this day . . . in 1346 artillery was first used in action by the English as Edward III led his army in the defeat of the French, led by Philip VI, at Crécy.

Born on this day were: Sir Robert Walpole (1676–1745), Britain's first prime minister and later the first Earl of Orford; Joseph Michel Montgolfier (1740–1810), the French inventor who (with his brother Jacques Étienne) built the first practical hot-air balloon; the French scientist Antoine Laurent Lavoisier (1743–94), who coined the word 'oxygen'; Prince Albert (1819–61), consort to Queen Victoria; Henry Fawcett (1833–84), the English political economist who in spite of being blinded as a young man became Postmaster-General in Gladstone's administration (in 1880) and introduced the parcel post and postal orders; the American inventor Lee De Forest (1873–1961), who was known as 'the father of radio'; John Buchan, Lord Tweedsmuir (1875–1940), the Scottish author and statesman, best remembered for his novel *The Thirty-Nine Steps*; Guillaume Apollinaire (1880–1918), the French poet who coined the word 'surrealism'; and the English novelist Christopher Isherwood (1904–86).

🐝 27 August

ON this day . . . in 1813 the Allies were defeated by Napoleon at the battle of Dresden.

Born on this day were: Georg Wilhelm Friedrich Hegel (1770–1831), the German philosopher; Charles Stewart Rolls (1877–1910), the founder (with Henry Royce) of Rolls Royce Ltd. and one of the pioneers of aviation in Great Britain; Samuel Goldwyn (1882–1974), the Polish-born American film producer; the English composer Eric Coates (1886–1957), whose best-known composition 'The Dam Busters' march (from his film score) was written a few years before his death; C(ecil) S(cott) Forester (1899–1966), the English novelist who created 'Captain Horatio Hornblower, RN'; Sir Donald Bradman (1908–), the first Australian cricketer to be knighted (in 1949) and the only Australian to have scored 100 First-Class centuries; Lyndon Baines Johnson (1908–73), the US president who succeeded to office with the assassination of President Kennedy; Lester Young (1909–59), the American jazz saxophonist known as 'the President', or 'Prez'; and Mother Teresa (1910–), the Yugoslavian-born missionary and recipient of the 1979 Nobel peace prize.

🐝 28 August

ON this day . . . in 1914 three German cruisers were sunk by ships of the Royal Navy in the battle of Heligoland Bight, the first major naval battle of World War I.

Born on this day were: Johann Wolfgang von Goethe (1749–1832), the German poet, dramatist, and scientist whose masterpiece, the drama *Faust*, occupied more than fifty years of his life; Sir Edward Burne-Jones (1833–98), the British painter and one of the first Pre-Raphaelite artists; Ira David Sankey (1840–1908), the American evangelist and hymn-writer who joined the preacher Dwight Lyman Moody in two tours of

Great Britain; and Sir John Betjeman (1906–84), the Poet Laureate 1972–84.

✌ 29 *August*

ON this day . . . in 1882 Australia defeated England at cricket in England for the first time; and the now famous 'obituary' notice was written to the *Sporting Times*: 'In affectionate remembrance of English Cricket which died at The Oval, 29th August, 1882. Deeply lamented by a large circle of sorrowing friends and acquaintances. R.I.P. NB. The body will be cremated and the ashes taken to Australia.'

Born on this day were: the English philosopher John Locke (1632–1704); the French painter Jean Auguste Dominique Ingres (1780–1867); Count Maurice Maeterlinck (1862–1949), the Belgian poet and dramatist whose play *Pelléas et Mélisande* was adapted as an opera by his friend Debussy; Jack Butler Yeats (1871–1957), the Irish landscape painter who was the brother of the poet W. B. Yeats; the Swedish actress Ingrid Bergman (1915–82); Charlie Parker (1920–55), the American jazz saxophonist and pioneer of modern jazz; and the British actor Sir Richard Attenborough (1923–   ).

✌ 30 *August*

ON this day . . . in 1860 the first British tramway was inaugurated at Birkenhead by an American, George Francis Train.

In 1862, at the second battle of Bull Run, in Virginia, the Union army was again defeated by the confederates, with 'Stonewall' Jackson (who had acquired his nickname at the first battle, in the previous year) playing a major part in the conflict.

Ernest Rutherford, first Baron Rutherford of Nelson (1871–1937), the New Zealand scientist who was the first to split the atom, was born on this day; and Sir Joseph John Thomson, the British scientist who discovered the electron and was a tutor of Rutherford at Cambridge, died in 1940. (Both men were recipients of the Nobel prize and were buried in Westminster Abbey.)

Also born on this day were: Jacques Louis David (1748–1825), court painter to Louis XVI and one of the deputies who voted for his execution; and Mary Wollstonecraft Shelley (1797–1851), wife of the poet Percy Bysshe Shelley and author of the novel *Frankenstein*.

ༀ 31 *August*

On this day . . . in 1942, at the battle of Alam al-Halfa, the German offensive led by General Rommel was halted by the British 8th Army under its new commander General Montgomery. (This engagement marked the turning-point in the North African Campaign.)

Born on this day were: Herman von Helmholtz (1821–94), the German scientist who invented the ophthalmoscope and originated the concept of conservation of energy; Maria Montessori (1870–1952), the Italian doctor who originated the method of education named after her, and the first woman in Italy to qualify as a doctor of medicine; DuBose Heyward (1885–1940), the American author whose novel *Porgy* was adapted as the libretto for Gershwin's opera *Porgy and Bess*; the American author William Saroyan (1908–81); and the American songwriter Alan Jay Lerner (1918–86).

John Bunyan, the English preacher who wrote *The Pilgrim's Progress*, died in 1688. ('So he passed over, and all the trumpets sounded for him on the other side.')

SEPTEMBER ❧

❧ 1 September

ON this day . . . in 1939 World War II began with the invasion of Poland by German forces.

Born on this day were: James John 'Gentleman Jim' Corbett (1866–1933), the American world heavyweight boxing champion 1892–7; and Rocky Marciano (1923–69), the American world heavyweight boxing champion who retired undefeated in 1956 after winning the title in 1952. (He was killed in an air crash the day before his 46th birthday.)

The French violinist Jacques Thibaud was killed in an air crash in 1953; and the English horn player Dennis Brain was killed in a car crash in 1957.

Also born on this day were: the German composer Engelbert Humperdinck (1854–1921); Sir Roger Casement (1864–1916), the Irish nationalist and British consular official who was hanged for treason; and Edgar Rice Burroughs (1875–1950), the American novelist who created 'Tarzan of the Apes'.

Nicholas Breakspear, the only Englishman to become pope (as Adrian IV), died in 1159. (His exact date of birth is not known.)

❧ 2 September

ON this day . . . in 1666 the Great Fire of London broke out in a baker's house in Pudding Lane, destroying some 13,000 houses and over eighty churches (including St Paul's) in the four days it lasted.

In 1752 the Julian calendar was used in Britain and the Colonies 'officially' for the last time, with the following day becoming 14 September in accordance with the Gregorian calendar and bringing Britain in line with the other European countries.

Born on this day were: John Howard (1726–90), the English philanthropist and prison reformer whose name is perpetuated through the Howard League for Penal Reform; Wilhelm Ostwald (1853–1932), the German chemist and recipient of the Nobel prize in 1909; and the British chemist Frederick Soddy (1877–1956), who named the isotope and was the recipient of the 1921 Nobel prize for chemistry.

In 1898 Kitchener's forces defeated the Khalifa's army at the battle of Omdurman and the young Winston Churchill took part in the cavalry charge of the 21st Lancers.

✌ 3 September

ON this day ... in 1939 Britain, New Zealand, Australia, and France declared war on Germany. (And the British prime minister Neville Chamberlain made his historic broadcast from No. 10 Downing Street at 11.15 a.m.: ' ... this country is at war with Germany ... It is the evil things we shall be fighting against—brute force, bad faith, injustice, oppression and persecution—and against them I am certain that right will prevail.')

On this same day the passenger liner *Athenia*, which was transporting women and children to the USA, was torpedoed and sunk by a U-boat with the loss of 112 lives.

Matthew Boulton (1728–1809), the English engineer and inventor whose partnership with James Watt advanced the development of the steam engine and originated the term 'horsepower', was born on this day.

In 1650 Cromwell defeated the Scots under David Leslie at the second battle of Dunbar; in 1651 Cromwell defeated the Scots under Charles II at the second battle of Worcester; and Cromwell died in 1658.

In 1916 the first Zeppelin was shot down over England.

✌ 4 September

ON this day ... in 1939 Canada and South Africa declared war on Germany, following the declarations of war by Britain, New Zealand, Australia, and France on the previous day.

Born on this day were: Vicomte François René de Chateaubriand (1768–1848), the French writer and statesman and noted gourmet who gave his name to the Chateaubriand steak; the German composer Anton Bruckner (1824–96); the French composer Darius Milhaud (1892–1974), who was one of the group of French composers known as 'Les Six'; the English novelist Mary Renault (1905–83); Dawn Fraser (1937–   ), the Australian swimmer who won eight Olympic medals, which is more than any other woman has won; and the American golfer Tom Watson (1949–   ), who was five times a winner of the British Open.

✻ 5 September

ON this day . . . in 1972, at the Olympic Games in Munich, Arab terrorists seized Israeli athletes as hostages. (Hostages, terrorists, and a German policeman were subsequently killed in a gun battle.)

Born on this day were: Louis XIV (1638–1715), the king of France known as 'the Sun King', whose reign (the longest of any European monarch) came to be described as the Augustan Age from its great up-surge in artistic and literary achievements; Johann Christian Bach (1735–82), the German composer who was the eleventh and youngest son of J. S. Bach and known as 'the London Bach' from his years as music master to the family of George III; John Wisden (1826–84), the English cricketer who (in 1864) founded *Wisden Cricketers' Almanack*, the 'Bible' of the game of cricket; Victorien Sardou (1831–1908), the French drama-tist whose play *La Tosca* inspired Puccini's famous opera; Jesse James (1847–82), the legendary American outlaw; the American film producer Darryl Zanuck (1902–79); and the American composer John Cage (1912–92).

✻ 6 September

ON this day . . . in 1620 101 English emigrants (to become known more than a hundred years afterwards as the Pilgrim Fathers) finally set sail from the Devon seaport of Plymouth in the *Mayflower* to found the colony of Plymouth in New England.

In 1901 the US president William McKinley was shot by an anarchist, Leon Czolgosz, and died eight days later; and in 1966 the South African prime minister Hendrik Verwoerd, who had survived an attempt on his life six years earlier, was stabbed to death in Parliament in Cape Town by a parliamentary messenger, Dmitri Tsafendas.

Born on this day were: John Dalton (1766–1844), the English chemist who first propounded the atomic theory; Sir H(enry) Walford Davies (1869–1941), the English composer and Master of the King's Music from 1934; and Billy Rose (1899–1966), the American impresario and songwriter.

7 September

ON this day . . . in 1776 the first submarine to be used in warfare, *The American Turtle*, made an unsuccessful attempt to attach a mine to Admiral Howe's flagship, HMS *Eagle*, in New York harbour.

In 1838 Grace Darling, the 22-year-old daughter of the Outer Farn lighthouse keeper, rowed out with her father to rescue survivors of a steamer wrecked off the Northumberland coast. (She was subsequently fêted as a national heroine.)

Born on this day were: Elizabeth I (1533–1603), the daughter of Henry VIII by his second wife Anne Boleyn; Samuel Wilberforce (1805–73), the Anglican bishop known as 'Soapy Sam' (from his apparently unctuous manner), who was a son of William Wilberforce, the slave-trade abolitionist; Sir Henry Campbell-Bannerman (1836–1908), the British Liberal prime minister 1905–8; Anna Mary 'Grandma' Moses (1860–1961), the American primitive painter who was in her late seventies when she began painting; Dame Edith Sitwell (1887–1965), the English poet and sister of the poets Osbert and Sacheverell Sitwell; and James Alfred Van Allen (1914–), the American scientist who discovered and gave his name to two belts of radiation surrounding the earth.

8 September

ON this day . . . in 1944 the first German long-range rocket, the V2 (from 'Vergeltungswaffe' or retaliation weapon), was launched from Holland and fell on London.

Born on this day were: Richard I 'Cœur de Lion' (1157–99), the third son of Henry II and king of England from 1189; the Italian poet Ludovico Ariosto (1474–1533); the French poet Frédéric Mistral (1830–1914), the recipient of the 1904 Nobel prize for literature; the Czech composer Antonín Dvořák (1841–1904), whose famous 'New World' Symphony was written in New York; W(illiam) W(ymark) Jacobs (1863–1943), the English short-story writer; Siegfried Sassoon (1886–1967), the English poet and novelist especially remembered for his *Memoirs* of World War I and his anti-war poems; Sir Harry Secombe (1921–   ), the British actor, comedian, and singer and one of the 'Goons' of the trend-setting radio comedy *Goon Shows* of the 1950s; Peter Sellers (1925–80), the British actor and comedian and another 'Goon'; the British playwright and novelist Michael Frayn (1933–   ); and the British composer Sir Peter Maxwell Davies (1934–   ).

৯৯ 9 *September*

On this day . . . in 1513 James IV of Scotland was killed at the battle of Flodden Field. ('The flowers of the forest are a' wede away'—Jane Elliot.)

In 1914 the first battle of the Marne was effectively ended with the retreat of the German forces in their advance on Paris; and in 1943, on the first day of the battle of Salerno, the Allied landing was met with fierce resistance from the German forces.

Born on this day were: Cardinal Richelieu (1585–1642), the French statesman and chief minister of Louis XIII; Luigi Galvani (1737–98), the Italian scientist who gave his name to 'galvanism'; Count Leo Nikolaevich Tolstoy (1828–1910), the Russian novelist and social reformer; and the English novelist James Hilton (1900–54), who created the imaginary paradise 'Shangri-La' in his novel *Lost Horizon*.

In 1087 William the Conqueror died after falling from his horse.

✌ 10 *September*

ON this day . . . in 1547 an English army under Edward Seymour, Duke of Somerset, defeated the Scottish army at the battle of Pinkie, near Edinburgh, with losses of some 200 against more than 10,000 on the Scottish side.

In 1919 new boundaries were settled at the Treaty of Saint-Germain which brought about the end of the Austrian Empire.

Born on this day were: Thomas Sydenham (1624–89), the physician known as 'the English Hippocrates'; the English architect Sir John Soane (1753–1837); Mungo Park (1771–1806), the Scottish explorer in Africa, where he died; Prince Ranjitsinhji (1872–1933), the England cricketer who became the Maharaja of Nawanagar; and the American golfer Arnold Palmer (1929–   ).

✌ 11 *September*

ON this day . . . in 1297 the Scottish patriot Sir William Wallace defeated the army of Edward I at the battle of Stirling Bridge.

In 1709 the Duke of Marlborough and Prince Eugene of Austria defeated the French Marshal Villars at the battle of Malplaquet, in France. (This was one of the bloodiest battles in the War of the Spanish Succession, with some 24,000 Allied losses against half that number for the French.)

Born on this day were: the French poet Pierre de Ronsard (1524–85); the Scottish poet James Thomson (1700–48), who wrote the words of the song 'Rule, Britannia!'; William Sidney Porter (1862–1910), the American short-story writer who used the pseudonym 'O. Henry'; and the English novelist and poet D(avid) H(erbert) Lawrence (1885–1930), whose last novel, *Lady Chatterley's Lover*, was not published in Great Britain in an unexpurgated edition until thirty years after his death, following the sensational court case of *Regina* v. *Penguin Books Ltd.* in 1960.

In 1917 Georges Guynemer, the legendary French fighter pilot, disappeared on active service. (It is said that French schoolchildren were told that 'he flew so high that he could not come down'.)

𝒫ᴀ 12 *September*

Oɴ this day ... in 1814 an intended attack on Baltimore, by the British fleet, was preceded with the bombardment of Fort McHenry, the vital point of the city's defences; and an American lawyer, Francis Scott Key, who witnessed the action, was inspired to write a poem, 'The Star-Spangled Banner', later to become the official national anthem of the USA.

Born on this day were: Richard Gordon Gatling (1818–1903), the American inventor of the first practical machine-gun; Herbert Henry Asquith, first Earl of Oxford and Asquith (1852–1928), the Liberal prime minister 1908–16; H(enry) L(ouis) Mencken (1880–1956), the American journalist; Maurice Chevalier (1888–1972), the French actor and singer; the British poet Louis MacNeice (1907–63); and James Cleveland 'Jesse' Owens (1913–80), the American athlete who, after winning three gold medals at the 1936 Olympic Games in Berlin, was snubbed by Hitler because he was coloured.

𝒫ᴀ 13 *September*

Oɴ this day ... in 1759 General James Wolfe was killed (and his opponent, the French Marquis de Montcalm, mortally wounded) at the battle of Quebec, which completed the British conquest of North America. ('Do they run already? Then I die happy.')

Born on this day were: Caspar Wistar (1761–1818), the American anatomist whose name was given to the genus *Wisteria*; Clara Schumann (1819–96), the German pianist and composer and the wife of the composer Robert Schumann; Walter Reed (1851–1902), the US Army surgeon who discovered the cause of yellow fever; the US Army general John Joseph Pershing (1860–1948), the commander-in-chief of the American Expeditionary Force, 1917–19; Franz von Hipper (1863–1932), the German commander at the naval battles of Dogger Bank and Jutland; Arthur Henderson (1863–1935), the British Labour politician and recipient of the 1934 Nobel peace prize; the Austrian composer Arnold Schoenberg (1874–1951), who invented the 'twelve note' system; Sir Robert Robinson (1886–1975), the British scientist and recipient of the

1947 Nobel prize for chemistry; J(ohn) B(oynton) Priestley (1894–1984), the British novelist and playwright; and Roald Dahl (1916–90), the British writer of short stories and books for children.

℘ 14 *September*

ON this day . . . in 1752 the Gregorian (or New Style) calendar was adopted in Great Britain and the preceding eleven days (3–13) were 'lost to history'.

In 1901 the US president William McKinley died, eight days after he was shot by an anarchist; and in 1911 the Russian premier Peter Stolypin was shot by an assassin at the theatre in Kiev, in the presence of the tsar and tsarina.

Born on this day were: Baron Alexander von Humboldt (1769–1859), the German naturalist and explorer; Lord Cecil of Chelwood (1864–1958), the British statesman and recipient of the 1937 Nobel peace prize; Margaret Sanger (1883–1966), the American pioneer of the birth control movement; and Sir Peter Scott (1909–89), the ornithologist and artist who was the son of the Antarctic explorer Robert Falcon Scott.

℘ 15 *September*

ON this day . . . in 1784 the first ascent in a hydrogen balloon in England was made by the Italian aeronaut Vincenzo Lunardi; and in 1830 the first accident caused by a moving train occurred at the opening of the Liverpool and Manchester Railway, when the British statesman William Huskisson was fatally injured by Stephenson's *Rocket*.

In 1916 the first tanks appeared in action in the third month of the battle of the Somme; and in 1940 the German air raids on Britain reached their climax and the anniversary has since been celebrated as 'Battle of Britain Day'. ('Never in the field of human conflict was so much owed by so many to so few', Winston Churchill.)

Born on this day were: the American novelist James Fenimore Cooper (1789–1851); the American president William Howard Taft (1857–1930); Dame Agatha Christie (1890–1976), the British novelist who created the fictional detectives 'Hercule Poirot' and 'Miss Marple'; Jean Renoir (1894–1979), the French film director whose father was the painter Pierre Auguste Renoir; Sir Donald Bailey (1901–85), the British engineer who invented and gave his name to the pre-fabricated bridge used extensively in World War II; and Umberto II (1904–83), the last king of Italy.

✄ 16 *September*

ON this day . . . in 1931 the last mutiny within the British Royal Navy came to an end as the Atlantic Fleet sailed from the Scottish seaport of Invergordon after resisting the order to put to sea on the previous day.

Born on this day were: Andrew Bonar Law (1858–1923), the British prime minister from October 1922 until May 1923, shortly before his death; the English poet Alfred Noyes (1880–1958); the French composer, conductor, and teacher Nadia Boulanger (1887–1979); Karl Doenitz (1891–1981), the German admiral and supreme commander of the Nazis in succession to Hitler, who was imprisoned for war crimes in 1946; and Sir Alexander Korda (1893–1956), the Hungarian-born film producer.

✄ 17 *September*

ON this day . . . in 1862 one of the decisive battles of the American Civil War ended with General McClellan repulsing General Lee's invasion of the North at Antietam; and in 1944 the battle of Arnhem began with an airborne invasion of Holland.

Born on this day were: Sir Francis Chichester (1901–72), the British yachtsman and aviation pioneer who made a solo circumnavigation of the world at the age of 65; Sir Frederick Ashton (1904–88), the English dancer and choreographer; John Creasey (1908–73), the English writer of crime fiction; and Stirling Moss (1929–), the former British champion racing driver.

৯১ 18 *September*

ON this day . . . in 1961 Dag Hammarskjöld, the US secretary-general, was killed in an air crash at the time when he was mediating in the Congo crisis. (He was subsequently awarded the 1961 Nobel peace prize.)

Born on this day were: Samuel Johnson (1709–84), the English lexicographer and 'great Cham of literature' (as the novelist Smollett described him); Jean Bernard Léon Foucault (1819–68), the French scientist who invented the gyroscope; the Australian composer Arthur Benjamin (1893–1960), best remembered for his *Jamaican Rumba*; the English actress Fay Compton (1894–1978), who was the sister of the novelist Sir Compton Mackenzie; and the Swedish-born actress Greta Garbo (1905–90).

৯১ 19 *September*

ON this day . . . in 1356 Edward, the Black Prince, defeated the French and took King John II as a prisoner at the battle of Poitiers.

In 1783 the Montgolfier brothers sent up the first balloon with live creatures on board—a sheep, a rooster, and a duck.

In 1881 the US president James Garfield died of wounds he had received from an assassin two months previously.

Born on this day were: Sir William Golding (1911–), the English novelist and recipient of the 1983 Nobel prize for literature; Emil Zatopek (1922–), the Czech Olympic athlete; and his wife Dana Zatopkova (1922–), who is also an Olympic athlete.

৯১ 20 *September*

ON this day . . . in 1792 the French defeated the Prussians at the battle of Valmy (and Goethe, who witnessed the conflict, wrote, 'This field and this day mark the beginning of a new epoch in the history of the world.'). In 1854 the first Victoria Crosses to be awarded to the British Army were won at Alma, in the first battle of the Crimean War.

Born on this day were: Sir James Dewar (1842–1923), the Scottish scientist who invented the vacuum flask (originally named after him as the Dewar-flask); Sir George Robey (1869–1954), the English actor and comedian known as 'the prime minister of mirth'; the American novelist and politician Upton Sinclair (1878–1968); the American jazz pianist and composer Ferdinand Joseph La Menthe 'Jelly Roll' Morton (1885–1941), who was one of the first to orchestrate jazz music as well as claiming to have invented jazz; the English poet Florence Margaret 'Stevie' Smith (1902–71); the British jazz musician and composer John Dankworth (1927–   ); and the Italian-born actress Sophia Loren (1934–   ).

✌ 21 *September*

ON this day . . . in 1327 Edward II was murdered in Berkeley Castle, in Gloucestershire, eight months after his enforced abdication.

Born on this day were: Richard Plantagenet, third Duke of York (1411–60), father of Edward IV and Richard III; Girolamo Savonarola (1452–98), the Italian political reformer and Dominican monk who was hanged and burned as a heretic; John Loudon McAdam (1756–1836), the Scottish engineer who invented and gave his name to macadamized roads; the English novelist and pioneer of science fiction H(erbert) G(eorge) Wells (1866–1946); the English composer Gustav Holst (1874–1934); Juan de la Cierva (1895–1936), the Spanish aeronautical engineer who invented the Autogiro; Learie Constantine (1902–71), the West Indian cricketer and statesman and the first cricketer to be honoured with a life peerage (as Lord Constantine of Maraval and Nelson); and Captain Charles Hazlitt Upham (1908–   ), the New Zealand soldier who was the only man to be awarded a bar to the Victoria Cross in World War II.

৯১ 22 September

On this day . . . in 1914 three British cruisers, HMS *Aboukir*, HMS *Hogue*, and HMS *Cressy*, were sunk by one German submarine, the U-9, in just over one hour. (This action alerted the British Admiralty to the deadly effectiveness of the submarine, which had been largely unrecognized up to that time.) In 1943 the German battleship *Tirpitz* was critically disabled by British midget submarines at her 'safe' anchorage in a Norwegian fiord.

Born on this day were: Anne of Cleves (1515–57), the fourth wife of Henry VIII; Michael Faraday (1791–1867), the English scientist who gave his name to the unit of electrical capacity, the farad, and was known as 'Sir Humphrey Davy's greatest discovery'; and the Welsh poet Dannie Abse (1923–   ).

In 1586 the soldier-poet Sir Philip Sidney was mortally wounded at the battle of Zutphen, in The Netherlands. (He is said to have given his water bottle to another dying soldier, saying, 'Thy necessity is yet greater than mine.') In 1776 Nathan Hale, the American soldier and hero of the War of Independence, was hanged by the British as a spy. ('I regret that I have but one life to give for my country.')

৯১ 23 September

On this day . . . in 1779, at the battle of Flamborough Head, John Paul Jones, the Scottish-born sailor (who had previously 'invaded' Britain at Whitehaven), was in command of a combined French and American fleet which captured the British ship *Serapis*. (Jones's ship, the *Bonhomme Richard*, was severely damaged in the action, but he refused to withdraw, proclaiming, 'I have not yet begun to fight.')

In 1803 Arthur Wellesley (later the Duke of Wellington) was victorious at the battle of Assaye (in India), his first major engagement as a commander and with a force composed largely of Indian troops.

In 1917 the legendary German fighter pilot Verner Voss was shot down and killed in an epic battle with five British fighter planes.

Born on this day were: Baroness Orczy (1865–1947), the Hungarian-born novelist and playwright who created 'the Scarlet Pimpernel'; John

Boyd Orr, first Baron Boyd Orr (1880–1971), the Scottish biologist and recipient of the 1949 Nobel peace prize; Walter Lippman (1889–1974), the American journalist and recipient of the Pulitzer prize for international reporting, in 1962; and the American film actor Mickey Rooney (1920–   ).

℘ 24 September

On this day... in 1852 the French engineer Henri Giffard made the first flight in an airship, which was powered by a steam engine of his own design.

In 1975 the British mountaineers Dougal Haston and Doug Scott became the first men to reach the summit of Mount Everest via the southwest face.

Born on this day were: Horace Walpole (1717–97), fourth Earl of Orford, the writer who was the fourth son of the British prime minister Sir Robert Walpole; the British writer and politician (Sir) A(lan) P(atrick) Herbert (1890–1971), who was often referred to simply as 'APH'; Francis Scott Key Fitzgerald (1896–1940), the American novelist whose first names were taken from the writer of the American national anthem, but who is best known as F. Scott Fitzgerald; and Sir Howard Walter Florey, later Baron Florey of Adelaide (1898–1968), the Australian scientist who shared the 1945 Nobel prize for medicine with Sir Alexander Fleming and Sir Ernst Chain for the discovery of penicillin.

In 1776 the oldest of the five 'classic' British horse-races, the St Leger (named after its founder Colonel Barry St Leger), was first run at Doncaster, in Yorkshire.

℘ 25 September

On this day... in 1066 Harold II, the king of England, defeated the king of Norway, Harald Hardrada, at the battle of Stamford Bridge, in Yorkshire. (Hardrada and Harold's brother Tostig, who had fought against him, were both killed in the battle.)

In 1915 British and French troops were sent into an attack on the German lines at Loos, on the first day of a battle which was to continue into October.

The English poet Samuel 'Hudibras' Butler died in poverty in 1680 (and a monument erected to his memory in Westminster Abbey carries the epigram, 'The poet's Fate is here in emblem shown: | He asked for Bread and he received a stone.').

Born on this day were: the English poet Felicia Dorothea Hemans (1793–1835), best remembered for her poem 'Casabianca', with the memorable first line, 'The boy stood on the burning deck...' (this relates how, at the battle of the Nile, the 13-year-old son of Louis Casabianca, the captain of the French flagship *L'Orient*, stayed with his father until 'There came a burst of thunder sound— | The boy—oh! where was he?'); the American novelist William Faulkner (1897–1962), the recipient of the 1949 Nobel prize for literature; and the Russian composer Dmitri Shostakovich (1906–75).

❧ 26 *September*

On this day... in 1580 Francis Drake and fifty sailors of the *Golden Hind* became the first Englishmen to circumnavigate the earth on their return to Plymouth after a voyage of more than thirty-three months.

In 1934 Cunard's largest passenger liner, the *Queen Mary*, was launched at Clydebank by Queen Mary.

Born on this day were: the French painter Jean Louis André Théodore Géricault (1791–1824); Ivan Petrovich Pavlov (1849–1936), the Russian physiologist and the first Russian to receive a Nobel prize, in 1904; the French pianist Alfred Cortot (1877–1962); Sir Barnes Wallis (1887–1979), the British aeronautical engineer especially remembered as the designer of the 'bouncing bombs' used in the 'Dam Busters' raid in 1943; the American-born British poet and playwright T(homas) S(tearns) Eliot (1888–1965); and the American composer and songwriter George Gershwin (1898–1937).

In 1953 sugar rationing in Great Britain was ended after almost fourteen years.

℘ 27 *September*

On this day... in 1825 George Stephenson drove his own steam locomotive on the opening run of the Stockton and Darlington Railway, the first public passenger railway.

In 1938 Queen Elizabeth, consort of George VI, launched the Cunard liner named after her (the largest passenger vessel at that time) at Clydebank; and in 1967 the *Queen Mary* completed her last transatlantic voyage.

Born on this day were: Samuel Adams (1722–1803), the American statesman and one of the signatories to the Declaration of Independence; George Cruikshank (1792–1878), the English caricaturist and illustrator; Louis Botha (1862–1919), the commander-in-chief of the Boer army and the first prime minister of South Africa; the English composer Cyril Scott (1879–1970); the French violinist Jacques Thibaud (1880–1953); the American song composer Vincent Youmans (1898–1946); the English poet Sir William Empson (1906–84); and Bernard Miles, Lord Miles of Blackfriars (1907–91), the English actor who founded the Mermaid Theatre, the first theatre opened in the City of London since the Restoration.

℘ 28 *September*

On this day ... in 1864 the First International was founded at a public meeting in London, when Karl Marx proposed a scheme for an International Workingmen's Association to some fifty delegates from various countries.

Born on this day were: John Jackson, the English pugilist known as 'Gentleman Jackson' (1769–1845), who was for eight years the national champion and a noted teacher of 'the noble art', with such as Lord Byron among his pupils and admirers ('And men unpractised in exchanging knocks | Must go to Jackson ere they dare to box.'); Richard Bright (1789–1858), the English physician who diagnosed and gave his name to Bright's disease; Prosper Mérimée (1803–70), the French writer whose novel *Carmen* was adapted for Bizet's opera of that name; Francis Turner Palgrave (1824–97), the English poet best remembered for his *Golden*

Treasury anthologies; the French statesman Georges Clemenceau (1841–1929); John French, Earl of Ypres (1852–1929), the commander-in-chief of the British army in France 1914–15; Kate Wiggin (1856–1923), the American author especially remembered for *Rebecca of Sunnybrook Farm*; Herman Cyril McNeile (1888–1937), the English writer who as 'Sapper' created 'Bulldog Drummond'; and the French film actress Brigitte Bardot (1934–   ).

ஃ 29 *September*

ON this day . . . in 1829 the first regular police patrols appeared on the streets of London. (The men became known as 'bobbies' from Sir Robert Peel, the Home Secretary, who had introduced a bill concerned with reorganizing London's police force.)

Born on this day were: Robert Clive, Lord Clive of Plassey (1725–74), the founder of the British Empire in India; Horatio Nelson, Lord Nelson (1758–1805), the British admiral who was immortalized by his death at the battle of Trafalgar; Mrs Elizabeth Cleghorn Gaskell (1810–65), the English novelist; and Enrico Fermi (1901–54), the Italian-born scientist who built the first American nuclear reactor and was the first Italian to win the Nobel physics prize outright (in 1938).

Richard II was deposed by his cousin Henry IV in 1399.

ஃ 30 *September*

ON this day . . . in 1938 Chamberlain, Daladier, Hitler, and Mussolini signed the Munich Agreement ('I believe it is peace for our time . . . peace with honour.').

Born on this day were: Lord Raglan (1788–1855), the British soldier who was the commander-in-chief of the British forces in the Crimean campaign; Lord Roberts, VC (1832–1914), the British soldier who was the commander-in-chief of the British forces in the Boer War; and Reinhard von Scheer (1863–1928), the German admiral who commanded the High Seas Fleet at the battle of Jutland.

Also born on this day were: the Irish composer Sir Charles Villiers Stanford (1852–1924); Hans Geiger (1882–1945), the German scientist who invented (with Lord Rutherford) the Geiger counter; the Russian violinist David Oistrakh (1908–74); Donald Swann (1923–   ), the British composer and performer; and the American author Truman Capote (1924–84).

OCTOBER &

❧ 1 October

IN 1938 German troops entered Czechoslovakia, precipitating the outbreak of World War II.

Born on this day were: Henry III (1207–72), the son of King John who succeeded his father in 1216; the third Duke of Grafton (1735–1811), the British prime minister who succeeded 'the elder Pitt' in 1767; Henry Clay Work (1832–84), the American songwriter who wrote 'Marching through Georgia'; the French composer Paul Dukas (1865–1935), who wrote the symphonic poem *The Sorcerer's Apprentice*; Stanley Holloway (1890–1982), the British actor, singer, and monologuist; Vladimir Horowitz (1904–89), the Ukrainian-born concert pianist who was the son-in-law of the conductor Arturo Toscanini; the thirty-ninth US president James Earl Carter (1924–); and the British-born actress and singer Julie Andrews (1935–).

❧ 2 October

ON this day ... in 1901 the British Royal Navy's first submarine, the American-designed Holland type, was launched. (The Irish-born designer John Philip Holland had originally intended his submarines to be used against the British and their development had been subsidized by the Fenians.)

Born on this day were: Paul von Hindenburg (1847–1934), the commander-in-chief of the German armies 1916–18; and Ferdinand Foch (1851–1929), the commander-in-chief of the Allied armies (as Marshal Foch) in 1918.

Also born on this day were: Richard III (1452–85); Sir William Ramsay (1852–1916), the first British scientist to receive the Nobel prize for chemistry, in 1904; Sir Patrick Geddes (1854–1932), the Scottish biologist and sociologist who coined the word 'conurbation'; Mahatma Gandhi (1869–1948), the Indian patriot and social reformer; Cordell Hull

(1871–1955), the American statesman and the longest-serving secretary of state (1933–44), who was the recipient of the Nobel peace prize in 1944; Julius Henry 'Groucho' Marx (1890–1977), the American comedian and foremost member of 'the Marx Brothers' comedy team; the British novelist and playwright Graham Greene (1904–91); and Robert Runcie (1921–   ), the 102nd archbishop of Canterbury.

ᜒᢀ 3 October

ON this day . . . in 1935 Abyssinia (now Ethiopia) was attacked by Italy.

In 1952 the first British atomic bomb was detonated on the Monte Bello Islands, off Western Australia.

Born on this day were: General Sir Samuel James Browne, VC (1824–1901), who gave his name to the Sam Browne, the sword or pistol belt for officers which he invented; the French painter Pierre Bonnard (1867–1947); the French novelist Henri Alain-Fournier (1886–1914), who was killed in action in World War I; Adolph Gysbert 'Sailor' Malan (1910–63), the South African fighter pilot ace of World War II; the British actor Sir Michael Hordern (1911–   ); and the American author Gore Vidal (1925–   ).

In 1990 the reunification of Germany was celebrated in Berlin in a three-day festival.

ᜒᢀ 4 October

ON this day . . . in 1957 *Sputnik I*, the first man-made earth satellite, was put into orbit from Russia; and exactly two years later *Lunik III* was launched, to take the first photographs of the back of the moon (which were transmitted to earth on 7 October 1959).

Born on this day were: Richard Cromwell (1626–1712), the son of Oliver Cromwell and Lord Protector from his father's death to his abdication in May 1659; the French painter Jean François Millet (1814–75); Rutherford Birchard Hayes (1822–93), the only US president to be elected through the voting of an electoral commission, in 1876;

Fred(eric) E(dward) Weatherly (1848–1929), the English songwriter whose numerous well-known lyrics included 'Danny Boy', 'Roses of Picardy', and 'The Holy City'; Sidney Paget (1860–1908), the English artist who illustrated the original 'Sherlock Holmes' stories; Lord Keyes of Zeebrugge (1872–1945), the British admiral who took his title from the raid he led on the Belgian seaport in 1918; Damon Runyon (1880–1946), the American short-story writer; the American film comedian and acrobat Joseph Francis 'Buster' Keaton (1896–1966); and the American film actor Charlton Heston (1924–   ).

৯ 5 October

ON this day . . . in 1914 a German reconnaissance aircraft was shot down in France and its pilot and observer were killed (they were the first airmen to die in an air battle), following an encounter with a French aircraft armed with a machine-gun; and in 1930 the British air minister was among the forty-seven killed when the R. 101, the world's largest dirigible at that time, crashed in France *en route* to India.

Born on this day were: Mary of Modena (1658–1718), the second wife of James, Duke of York, later James II; Chester Alan Arthur (1830–86), the US vice-president who succeeded to the presidency following the assassination of James Garfield in 1881; Louis Jean Lumière (1864–1948), the French pioneer of photography who (with his brother Auguste), invented the cinematograph; Václav Havel (1936–   ), the Czech playwright who became the last president of his country; and Bob Geldof (1954–   ), the Irish singer who was awarded an honorary knighthood for his organization of international famine relief.

In 1972 the United Reformed Church was formed with the union of the Congregational Church in England and Wales and the Presbyterian Church of England.

✇ 6 October

On this day . . . in 1536 William Tyndale, the English translator of the New Testament, was strangled and burned at the stake for heresy at Vilvorde, in Flanders.

In 1981 President Sadat of Egypt was assassinated by Muslim extremists at a military parade on the eighth anniversary of the Arab–Israeli War.

Born on this day were: Jenny Lind (1820–87), the internationally famous coloratura soprano known as 'the Swedish Nightingale'; the Polish composer and pianist Karol Szymanowski (1883–1937); Charles Édouard Jeanneret (1887–1965), the Swiss architect known as 'Le Corbusier' (who maintained that 'A house is a machine for living in'); and Thor Heyerdahl (1914–   ), the Norwegian anthropologist who led the *Kon-Tiki* expedition from Peru to a South Pacific island in 1947.

In 1927 the first successful 'talkie', the Warner Brothers' film *The Jazz Singer*, starring Al Jolson, was given its first public performance in New York.

✇ 7 October

On this day . . . in 1571 Don John of Austria, commanding the combined Spanish, Venetian, and Genoese forces, defeated the Turkish fleet at the battle of Lepanto, at the Gulf of Corinth. (With an estimated 33,000 men killed, this was probably the bloodiest naval battle of premodern times.)

Born on this day were: William Laud (1573–1645), the archbishop of Canterbury who was executed for treason; James Whitcombe Riley (1849–1916), the American writer known as the 'Hoosier poet' (he was born in Indiana, the 'Hoosier' state), who created 'Little Orfant Annie'; Niels Bohr (1885–1962), the Danish scientist and recipient of the 1922 Nobel prize in physics; the German Gestapo chief Heinrich Himmler (1900–45); and Shura Cherkassky (1911–   ), the Russian-born American pianist.

ஃ 8 October

ON this day . . . in 1912 the First Balkan War (against Turkey) began—from which World War I arose.

Born on this day were: Lord Rowton (1838–1903), the English philanthropist and politician (he was a private secretary to Disraeli) who originated and gave his name to 'poor men's hostels' in London, the Rowton Houses; John Milton Hay (1838–1905), the American statesman who was a private secretary to Lincoln and later secretary of state under McKinley and Theodore Roosevelt; John Cowper Powys (1872–1963), the best known of three brothers and writers of Welsh descent; Sir Alfred Munnings (1878–1959), the English painter (especially of horses) and president of the Royal Academy 1944–9; Eddie Rickenbacker (1890–1973), the American World War I fighter pilot who was credited with a record sixty-nine victories; Juan Perón (1895–1974), the Argentinian soldier, twice president of Argentina; and Betty Boothroyd (1929–   ), the first woman Speaker in the House of Commons (from 1992).

ஃ 9 October

ON this day . . . in 1470 Henry VI was restored to the throne of England after almost ten years (though some six months later he was again deposed and then murdered in the Tower of London).

In 1934 King Alexander I of Yugoslavia and the French foreign minister Jean Louis Barthou were assassinated in Marseilles by a Macedonian terrorist.

Born on this day were: the French composer Camille Saint-Saëns (1835–1921); the British comedy actor Alastair Sim (1900–76); Lord Hailsham (1907–   ), the former Lord Chancellor whose father was a Lord Chancellor; the French comedy actor (as 'Monsieur Hulot') and film director Jacques Tati (1908–82); Lord Coggan (1909–   ), the 101st archbishop of Canterbury; and John Lennon (1940–80), the English musician and songwriter and one of the 'Beatles', who was murdered in New York.

⚘ 10 *October*

On this day... in 1961 the whole population of the South Atlantic island of Tristan da Cunha was evacuated to Britain after a volcano erupted.

In 1973 the US vice-president Spiro Agnew resigned over his conviction for income tax evasion.

Born on this day were: the French painter Jean Antoine Watteau (1684–1721); Henry Cavendish (1731–1810), the English scientist who, among other things, discovered hydrogen; the Italian composer Giuseppe Verdi (1813–1901); Stephanus Johannes Paulus Kruger (1825–1904), the South African statesman whose name is perpetuated through the 'krugerrand'; Lord Nuffield (1877–1963), the British motor car manufacturer who as William Richard Morris pioneered the mass production of cheap cars in Britain; the Russian-born composer Vladimir Dukelsky (1903–69), who as 'Vernon Duke' was a successful writer of popular songs; the American songwriter Johnny Green (1908–89); and the British playwright Harold Pinter (1930–   ).

⚘ 11 *October*

On this day... in 1899 the Boer War began.

Born on this day were: Sir George Williams (1821–1905), the English social reformer who founded the YMCA; Eleanor Roosevelt (1884–1962), the wife of the thirty-second US president and a niece of the twenty-sixth president, but a politician in her own right; François Mauriac (1885–1970), the French writer and recipient of the 1952 Nobel prize for literature; the English writer Ethel Mannin (1900–84); Bobby Charlton (1937–   ), the England footballer and the second to win 100 international caps; and the Brazilian tennis player Maria Bueno (1939–   ) who was three times the Wimbledon women's singles champion.

In 1531 the Swiss reformer Ulrich Zwingli died at the battle of Kappel, which virtually ended the Swiss religious wars; and in 1797 Admiral Duncan led the British fleet to victory over the Dutch at the battle of Camperdown, off the north coast of Holland.

In 1982 the wreck of Henry VIII's flagship, the *Mary Rose*, was raised from the seabed off Portsmouth.

✺ 12 *October*

ON this day . . . in 1492 Christopher Columbus is believed to have discovered America, and this day is celebrated as Columbus Day in the USA.

In 1654 the Dutch painter Carel Fabritius, one of Rembrandt's best-known pupils and a teacher of Vermeer, was one of the many victims of a great explosion in Delft (as he was working on a portrait); and in 1984 the IRA exploded a bomb in the Grand Hotel in Brighton, which was intended to kill the British prime minister and other members of the cabinet attending the Conservative Party conference.

Born on this day were: Edward VI (1537–53), the only son of Henry VIII by his third wife Jane Seymour; Isaac Newton Lewis (1858–1931), the American army officer who invented the machine-gun which was named after him; Elmer Ambrose Sperry (1860–1930), the American inventor of the gyrocompass; James Ramsay MacDonald (1866–1937), Britain's first Labour prime minister; the English composer Ralph Vaughan Williams (1872–1958); and the Italian tenor Luciano Pavarotti (1935–   ).

In 1915 an English nurse, Edith Cavell, was executed by a German firing squad for assisting the escape of Allied soldiers through Belgium. ('I realize that patriotism is not enough. I must have no hatred or bitterness towards anyone.')

✺ 13 *October*

ON this day . . . in 1812 General Sir Isaac Brock, the British soldier known as 'the hero of Upper Canada', was killed at the battle of Queenston Heights. (This action effectively prevented any further invasion of Canada by the Americans.)

In 1815 Joachim Murat, the French marshal who married Napoleon's sister Caroline and became king of Naples, was executed by a firing squad after a court martial.

Born on this day were: Rudolf Virchow (1821–1902), the German scientist and politician known as 'the founder of modern pathology'; Lillie Langtry (1852–1929), the actress known as 'the Jersey Lily' (from her birthplace in the Channel Islands), who was a mistress of Edward VII; the American jazz pianist Art Tatum (1910–56), generally regarded as the greatest exponent of jazz at the keyboard; and Baroness Thatcher of Kesteven in the county of Lincolnshire (1925–   ), Britain's first woman prime minister (in 1979) and, as Mrs Margaret Thatcher, the longest-serving British prime minister in the twentieth century.

৪৯ 14 October

ON this day... in 1066 the English king Harold II was defeated and killed by the invading army of William, Duke of Normandy, at the battle of Hastings (which in fact was fought some five miles out of the town at Senlac Hill).

In 1806 the French Marshal Louis Davout defeated one Prussian army at the battle of Auerstädt as Napoleon defeated another at the battle of Jena.

Born on this day were: Sir Peter Lely (1618–80), the Dutch-born portrait painter who was employed by Charles I, Oliver Cromwell, and Charles II; James II (1633–1701), the second son of Charles I and younger brother of Charles II, whom he succeeded as king of England in 1685; William Penn (1644–1718), the English Quaker who founded Pennsylvania; George Grenville (1712–70), the British prime minister 1763–5; Eamon de Valera (1882–1975), the American-born president of the Irish Republic 1959–73; the New Zealand-born writer Katherine Mansfield (1888–1923); Dwight D(avid) Eisenhower (1890–1969), the American general and supreme Allied commander in World War II who became the thirty-fourth US president in 1952; the American poet Edward Estlin Cummings (1894–1962), who was known as 'e. e. cummings'; and the British singer Harry Webb (1940–   ), known as Cliff Richard.

✣ 15 October

ON this day ... in 1582 the Gregorian (or New Style) calendar was adopted in Italy, France, Spain, and Portugal; and the preceding ten days (5–14th) were 'lost to history'.

In 1917 the Dutch dancer 'Mata Hari' was shot by the French for spying for the Germans; and in 1945 Pierre Laval, a former prime minister of France, was shot for collaborating with the Germans.

Born on this day were: the Roman poet Virgil (70–19 BC); Evangelista Torricelli (1608–47), the Italian physicist and mathematician who invented the barometer and was Galileo's amanuensis; Friedrich Wilhelm Nietzsche (1844–1900), the German philosopher whose idea of the 'superman' inspired the Nazi movement; Marie Stopes (1880–1958), the English palaeobotanist and pioneer of birth control; Sir P(elham) G(renville) Wodehouse (1881–1975), the English novelist and songwriter; the English novelist and scientist Lord C(harles) P(ercy) Snow (1905–80); and the American novelist Mario Puzo (1920–   ).

✣ 16 October

ON this day ... in 1555 the Protestant martyrs Bishop Hugh Latimer and Bishop Nicholas Ridley were burned at the stake for heresy, opposite Balliol College in Oxford. ('Be of good comfort, Master Ridley, and play the man. We shall this day light such a candle by God's grace in England, as I trust never shall be put out.')

In 1793 Marie Antoinette was guillotined in Paris. ('Farewell, my children, for ever. I am going to your father.')

In 1946 ten Nazi war criminals (Franck, Frick, Jodl, Kaltenbrunner, Keitel, Ribbentrop, Rosenberg, Sauckel, Streicher, and Seyss-Inquart) were hanged at Nuremberg. (An eleventh, Hermann Goering, committed suicide a few hours before the time fixed for his execution.)

Born on this day were: the American lexicographer Noah Webster (1758–1843), who gave his name to the first complete American dictionary; Lord Cardigan (1797–1868), famed for leading the charge of the Light Brigade at the battle of Balaclava and for the knitted jacket named after him; Major Walter Clopton Wingfield (1833–1912), the English army officer who claimed to have invented lawn tennis; Oscar Wilde (1854–1900), the Irish playwright and poet; David Ben-Gurion (1886–1973), Israel's first prime minister; the American playwright Eugene O'Neill (1888–1953); and the German writer Günter Grass (1927–   ).

In 1908 the American-born aviator and inventor 'Colonel' Samuel Franklin Cody made the first officially recognized powered flight in Britain.

✤ 17 October

ON this day . . . in 1777 the battle of Saratoga ended with the surrender of the British commander General Burgoyne to General Gates. (This engagement marked the turning-point in the American War of Independence.)

Born on this day were: John Wilkes (1727–97), the English political reformer who secured the freedom of the press to report parliamentary debates; Georg Büchner (1813–37), the German poet whose play *Woyzeck* was adapted as the opera *Wozzeck* by Alban Berg; the British novelist Elinor Glyn (1864–1943); Herbert Howells (1892–1983), the English composer whose music was played at both the 1937 and the 1953 coronations; and the American playwright Arthur Miller (1915–   ).

✤ 18 October

ON this day . . . in 1967 a Russian spacecraft made the first soft landing on Venus.

Born on this day were: Richard Nash (1674–1762), the British socialite and gambler known as 'Beau Nash' and 'the King of Bath'; the Italian painter Antonio Canaletto (1697–1768); Thomas Love Peacock (1785–1866), the English poet and novelist; Christian Friedrich Schonbein (1799–1868), the German scientist who discovered ozone; Henri Bergson (1859–1941), the French philosopher who won the Nobel prize for literature in 1927; and the Czech-born tennis player Martina Navratilova (1956–   ), who won the Wimbledon women's singles a record nine times.

In 1865 the British prime minister Lord Palmerston died in office, two days short of his 81st birthday, saying (it is recorded), 'Die, my dear Doctor, that's the *last* thing I shall do!'

✇ 19 October

ON this day . . . in 1781 at Yorktown, Virginia, the last major battle of the American War of Independence ended with the surrender of the British commander General Cornwallis to General Washington; and in 1864, at the battle of Cedar Creek, the Unionist commander turned defeat into a famous victory after riding twenty miles to the scene of the battle. (His feat was commemorated in Thomas Buchanan Read's poem 'Sheridan's Ride': 'The terrible grumble, and rumble, and roar, | Telling the battle was on once more, | And Sheridan twenty miles away.')

Born on this day were: Sir Thomas Browne (1605–82), the English physician and writer ('We all labour against our own cure; for death is the cure of all diseases.') who died on the seventy-seventh anniversary of his birthday; Leigh Hunt (1784–1859), the English poet and essayist depicted as 'Skimpole' in Dickens's *Bleak House*; Tom Taylor (1817–80), the English dramatist whose play *Our American Cousin* was being performed at Ford's Theater, Washington, as Lincoln was assassinated; and August Marie Louis Nicolas Lumière (1862–1954), the French chemist who, with his brother Louis Jean, invented an early motion-picture camera and a process of colour photography.

In 1813 Napoleon was defeated at Leipzig by the Allies in a bloody engagement known as 'the battle of the Nations', in which some 500,000 troops were involved.

✇ 20 October

ON this day . . . in 1805 the Austrian general Karl Mack surrendered to Napoleon's army at the battle of Ulm; and in 1827 the Turkish fleet was destroyed by the Allied squadrons of Britain, France, and Russia at the battle of Navarino.

Born on this day were: Sir Christopher Wren (1632–1723), the English architect who rebuilt St Paul's and more than fifty other City of London churches after the Great Fire; Lord Palmerston (1784–1865), the British prime minister 1855–8 and 1859–65; Sir Colin Campbell (1792–1863), the British soldier who was appointed by Palmerston as commander-in-chief in India at the outbreak of the Mutiny (after which he was created Lord Clyde); Thomas Hughes (1822–96), the English writer and reformer who wrote *Tom Brown's Schooldays*; the French poet Arthur Rimbaud (1854–91); the American philosopher John Dewey (1859–1952); the American composer Charles Ives (1874–1954); Sir James Chadwick (1891–1974), the English scientist who was awarded the Nobel prize for physics in 1935 for his discovery of the neutron; and the British actress Dame Anna Neagle (1904–86).

✣ 21 *October*

ON this day . . . in 1805 Lord Nelson was mortally wounded as the French fleet commanded by Villeneuve was defeated at the battle of Trafalgar; and in 1960 the Royal Navy's first nuclear submarine, the *Dreadnought*, was launched by Queen Elizabeth at Barrow, where in 1901 the Royal Navy's first submarine had been launched.

Born on this day were: the English poet Samuel Taylor Coleridge (1772–1834); Alfred Nobel (1833–96), the Swedish inventor who discovered and manufactured dynamite and founded the Nobel prizes; Sir Georg Solti (1912–   ), the Hungarian-born conductor; John Birks 'Dizzy' Gillespie (1917–93), the American trumpet player and pioneer of modern jazz; and the English trumpet player and composer Sir Malcolm Arnold (1921–   ).

✣ 22 *October*

ON this day . . . in 1797 the first successful parachute descent was made by a Frenchman, André-Jacques Garnerin, who fell from a balloon at some 2,200 feet over Paris; and in 1909 the French aviatrix Élise Deroche became the first woman to make a solo flight.

Born on this day were: the Hungarian composer and pianist Franz Liszt (1811–86); Lord Alfred Douglas (1870–1945), the poet and son of the eighth Marquis of Queensberry and associate of Oscar Wilde; the English artist Sir Matthew Smith (1879–1959); and the South African novelist Doris Lessing (1919–   ).

✌ 23 October

On this day . . . in 1642 the royalists under Charles I opposed the parliamentarians under the Earl of Essex at the battle of Edgehill in Warwickshire, the first major encounter of the Civil War, which ended indecisively; in 1942 the British general Montgomery began his first major offensive against the German general Rommel at the battle of El Alamein in Egypt; and the biggest naval engagement of World War II began, between the Americans and the Japanese, at the battle of the Philippine Sea, in 1944.

Born on this day were: Pierre Larousse (1817–75), the French lexicographer and encyclopaedist; Robert Bridges (1844–1930), the Poet Laureate who succeeded Alfred Austin; Douglas Jardine (1900–58), the England cricket captain who led the England team in the controversial 'bodyline' tour of Australia in 1932–3; and the Brazilian footballer Pelé (Edson Arantes do Nascimento) (1940–   ).

✌ 24 October

On this day . . . in 1945 Vidkun Quisling, the Norwegian politician who collaborated in the German occupation of Norway, was executed by a firing squad.

Born on this day were: Anton von Leeuwenhoek (1632–1723), the Dutch scientist who discovered the existence of bacteria; Sarah Hale (1788–1879), the American writer whose *Poems for Our Children* included 'Mary had a little lamb' (the first words to be recorded in sound); the English actress Dame Sybil Thorndike (1882–1976); and the English actor Jack Warner (1894–1981).

The United Nations formally came into existence in 1945.

❧ 25 October

ON this day ... in 1415 Henry V led his army against the French at Agincourt ('This day is called the feast of Crispian'—William Shakespeare); and in 1854 Lord Cardigan led the Light Brigade in the abortive charge against the Russian guns at Balaclava ('Into the valley of Death | Rode the six hundred'—Alfred, Lord Tennyson).

In 1944, at the battle of Leyte Gulf, the main fleet of Japan was defeated by the US 3rd and 7th fleets in an engagement (also known as the battle of the Philippine Sea) which saw the end of the conventional battleship and Japanese sea power.

Born on this day were: Lord Grenville (1759–1834), the British prime minister who headed the 'All the Talents' administration (1806–7) and whose father had been prime minister; Lord Macaulay (1800–59), the historian; the English painter Richard Parkes Bonington (1802–28); the Austrian composer and conductor Johann Strauss the younger (1825–99), best remembered for his waltz *The Blue Danube*; the French composer Georges Bizet (1838–75), best remembered for his opera *Carmen*; the Spanish painter Pablo Picasso (1881–1973); and Richard Evelyn Byrd (1888–1957), the American polar explorer and the first man to fly over the North Pole (in 1926) and the South Pole (in 1929).

The English poet Geoffrey Chaucer died in 1400. (His exact date of birth is not known.)

In 1881 Wyatt Earp, the marshal of Tombstone, Arizona, with his two brothers and the 'professional' killer 'Doc' Holliday, challenged the Clanton gang and the McLowerys in the legendary 'Gunfight at the OK Corral', in which the McLowerys and one of the Clantons were killed. (This incident is regarded as a milestone in law enforcement in the 'Wild West'.)

❧ 26 October

ON this day ... in 1942 the combined Japanese fleet was engaged by American carrier groups at the battle of Santa Cruz Islands, with both sides sustaining heavy losses.

Born on this day were: the Italian composer Domenico Scarlatti (1685–1757), who was the son of the composer Alessandro Scarlatti; Johan Helmich Roman (1694–1758), the composer known as 'the father of Swedish music'; the French Revolutionary leader Georges Jacques Danton (1759–94); Joseph Aloysius Hansom (1803–82), the English architect who invented the hansom cab; C(harles) P(restwich) Scott (1846–1932), the English journalist and politician and editor of the *Manchester Guardian* 1872–1929; and François Mitterrand (1916–), the president of France from 1981.

℘ 27 October

On this day . . . in 1553 the Spanish theologian and physician Michael Servetus was burned at the stake at Geneva by order of John Calvin. (His last recorded words were, 'Christ, Son of the eternal God, have mercy upon me,' in spite of Calvin's insistence on his saying, 'the eternal Son of God'.)

Born on this day were: James Cook (1728–79), the English mariner and explorer who discovered and named Botany Bay, in Australia (from the abundance of plants he found there); the Italian composer and virtuoso violinist Niccolò Paganini (1782–1840); Theodore Roosevelt (1858–1919), the twenty-sixth US president, who gave his name to the 'teddy-bear'; the Welsh poet Dylan Thomas (1914–53); and the American poet Sylvia Plath (1932–63), who was married to the English poet Ted Hughes.

℘ 28 October

On this day . . . in 1916 Oswald Boelcke, the German fighter pilot and pioneer of air combat, was killed accidentally on a routine flight; and in 1949 the French violinist Ginette Neveu and Marcel Cerdan, the French boxing champion, were among the forty-eight passengers and crew who were killed in a plane crash in the Azores.

Born on this day were: Cornelis Jansen (1585–1638), the Dutch theologian who founded and gave his name to a sect within the Catholic Church; the English novelist Evelyn Waugh (1903–66); the British artist

Francis Bacon (1909–92); Jonas Edward Salk (1914–   ), the American scientist who developed and gave his name to a vaccine against poliomyelitis; and the British jazz singer Cleo Laine (1927–   ).

✄ 29 October

ON this day ... in 1618 Sir Walter Ralegh was beheaded at Whitehall under a sentence brought against him fifteen years earlier for conspiracy against James I. (On being asked which way he wished to lay his head on the block, he replied: 'So the heart be right, it is no matter which way the head lies.')

Born on this day were: James Boswell (1740–95), the Scottish author and personal biographer of Dr Johnson; Wilfred Rhodes (1877–1973), the Yorkshire and England cricketer who was the oldest man to play in a Test match (at the age of 52 in 1930); Fanny Brice (1891–1951), the American entertainer whose life was celebrated in four different films from 1933 to 1975; Joseph Goebbels (1897–1945), Hitler's propaganda minister; and Vivian Ellis (1904–   ), the English composer and author.

In 1929 the US stock market collapsed and this day became known as 'Black Tuesday', or the day of 'the Wall Street Crash'.

✄ 30 October

ON this day ... in 1925 the Scottish inventor John Logie Baird made the first televised transmission of a moving image from an attic in Frith Street, London W1. (The first person to appear on a television screen was in fact a 15-year-old office boy, who received a payment of two shillings and sixpence from Baird.)

Born on this day were: John Adams (1735–1826), the first vice-president of the USA and the second president, 1797–1801; the Irish dramatist and politician Richard Brinsley Sheridan (1751–1816); Viscount Goderich (1782–1859), the British prime minister 1827–8; Adelaide Anne Proctor (1825–64), the English poet mainly remembered for her poem 'The Lost Chord', which was set to music by Sullivan; the American poet Ezra Pound (1885–1972); and Philip Heseltine (1894–

1930), the English composer and musicologist who used the pseudonym 'Peter Warlock'.

℘ 31 *October*

ON this day ... in 1517 Martin Luther began his attack on the papal system by nailing his ninety-five theses on indulgences to the church door at Wittenberg, denying to the pope the right to forgive sins.

In 1952 the US exploded the first hydrogen bomb at Eniwetok Atoll in the Pacific; in 1956 the 'Suez Adventure' began with the bombing of Egyptian military installations by an Anglo-French force; and in 1984 Mrs Indira Gandhi, the Indian prime minister, was assassinated.

Born on this day were: John Evelyn (1620–1706), the English writer famed for his diaries; John Keats (1795–1821), the English poet whose death at the age of 25 was lamented by Shelley in his poem 'Adonais'; Sir Joseph Wilson Swan (1828–1914), the English scientist whose invention of the incandescent lamp occurred independently of Edison's invention; Cosmo Gordon Lang (1864–1945), the ninety-seventh archbishop of Canterbury 1928–42; Marie Laurencin (1885–1956), the French painter who was nicknamed 'the Fauvette'; Chiang Kai-shek (1887–1975), the Chinese general and chairman of the National Republic of China 1943–9; and Michael Collins (1930–   ), the American astronaut who piloted the command module in the first moon landing in 1969.

NOVEMBER ❧

❧ 1 November

On this day ... in 1914 the British admiral Sir Christopher Cradock went down with his flagship, HMS *Good Hope*, at the battle of Coronel, after engaging an overwhelmingly stronger squadron commanded by the German admiral Maximilian von Spee.

Born on this day were: Spencer Perceval (1762–1812), the only British prime minister to have been assassinated; William Whiting (1825–78), the English hymn-writer best remembered for his hymn 'for those at sea' ('Eternal Father, strong to save . . .'); Stephen Crane (1871–1900), the American novelist and war correspondent best remembered for *The Red Badge of Courage*; the English composer Roger Quilter (1877–1953); the English painter L(aurence) S(tephen) Lowry (1887–1976); Alexander Alekhine (1892–1946), the Russian chess master and world champion who became a French citizen; the English poet Edmund Blunden (1896–1974); Naomi Mitchison (1897–), the British writer who is the daughter of the scientist John Scott Haldane; and the Spanish soprano Victoria de los Angeles (1923–).

❧ 2 November

On this day ... in 1947 the American millionaire and aircraft designer Howard Hughes piloted his H.4 Hercules flying-boat in a test flight off Long Beach Harbor, California. (This was the first and last flight of the massive eight-engined aircraft with its record 97.51 m. wing span.)

Born on this day were: Edward V (1470–83), king of England from 9 April to 25 June 1483, who was murdered in the Tower with his brother, the young Duke of York (probably with the connivance of his uncle Richard, Duke of Gloucester, who had assumed the crown as Richard III); Marie Antoinette (1755–93), the queen of France who followed her husband (Louis XVI) to the guillotine after being tried by the

Revolutionary Tribunal; James Knox Polk (1795–1849), the eleventh US president; Warren Gamaliel Harding (1865–1923), the twenty-ninth US president, who died in office in the second year of his presidency; Aga Khan III (1877–1957), the hereditary head of the Ismailian Muslims who was the president of the League of Nations (in 1937) and the owner of five Derby winners; and the American film actor Burt Lancaster (1913–   ).

ಳಿ 3 November

On this day . . . in 1871 Henry Morton Stanley met David Livingstone at Ujiji. ('Dr Livingstone, I presume?')

In 1957, the Russian artificial satellite *Sputnik II* was launched with a dog, 'Laika', on board, which was the first living creature to travel in space.

Born on this day were: Henry Ireton (1611–51), the parliamentarian commander and son-in-law of Oliver Cromwell who was a signatory of the warrant for the execution of Charles I; John Montague, the fourth Earl of Sandwich (1718–92), who as First Lord of the Admiralty gave his name (through Captain Cook) to the Sandwich Islands and (through his addiction to the gaming table) the 'sandwich'; Daniel Rutherford (1749–1819), the Scottish scientist who discovered nitrogen; the German publisher Karl Baedeker (1801–59), who gave his name to the famous guidebooks; the Italian composer Vincenzo Bellini (1801–35); and the American world heavyweight boxing champion Larry Holmes (1949–   ).

ಳಿ 4 November

On this day . . . in 1922 the British archaeologist Howard Carter discovered the steps leading to the tomb of the Egyptian pharaoh Tutankhamun.

Born on this day were: Francois Le Clerc du Tremblay (1577–1638), the confidant to Cardinal Richelieu known as Père Joseph or 'l'Éminence Grise' (from the colour of his habit and the grey or shadowy power he exercised over Richelieu); William III (1650–1702), the king of England, Scotland, and Ireland who reigned jointly with Mary II, the daughter of James II by Anne Hyde, in succession to James II; Augustus Montague Toplady (1740–78), the English clergyman who wrote 'Rock of Ages' and other well-known hymns; the English novelist and playwright Eden Phillpotts (1862–1960); the English composer Herman Finck (1872–1939); and Will Rogers (1879–1935), the American actor and writer known as 'the cowboy philosopher'.

In 1918 the English poet Wilfred Owen was killed in action on the Western Front, just one week before the Armistice. ('Red lips are not so red | As the stained stones kissed by the English dead.')

✃ 5 November

On this day . . . in 1605 Guy Fawkes was betrayed and arrested in an attempt to blow up the Houses of Parliament ('the Gunpowder Plot').

Born on this day were: Ella Wheeler Wilcox (1850–1919), the American poet best remembered for her lines 'Laugh and the world laughs with you; | Weep, and you weep alone . . .'; the English poet James Elroy Flecker (1884–1915), best remembered for 'The Golden Journey to Samarkand'; John Burdon Sanderson Haldane (1892–1964), the Scottish scientist and son of John Scott Haldane; Vivien Leigh (1913–67), the British actress and second wife of Laurence Olivier; and the British jockey Lester Piggott (1935–   ), who has ridden a record nine Derby winners.

In 1854 nineteen Victoria Crosses were won in the defeat of the Russians at the battle of Inkerman; and in 1940 Captain Fegen of the armed merchant cruiser HMS *Jervis Bay* won a posthumous VC after engaging the German battleship *Admiral Scheer* (sister ship of the *Bismarck*) in an attempt to protect a convoy of thirty-seven merchant ships. (The *Jervis Bay* was predictably out-gunned by the *Scheer* and Captain Fegen went down with his ship, but the majority of the convoy escaped.)

✋ 6 November

BORN on this day were: Colley Cibber (1671–1757), the English actor and dramatist and Poet Laureate 1730–57; Alois Senefelder (1771–1834), the German inventor of lithography; Adolphe Sax (1814–94), the Belgian inventor of the saxophone; Charles Henry Dow (1851–1902), the American financial journalist who (with Edward D. Jones) inaugurated and gave his name to the 'Dow–Jones' averages; John Philip Sousa (1854–1932), the American bandmaster known as 'the March King' who gave his name to the sousaphone; and Sir John Alcock (1892–1919), the pilot of the first aircraft to make a non-stop transatlantic flight.

In 1893 the Russian composer Peter Ilich Tchaikovsky died, after being persuaded to take arsenic to avoid revelations of his alleged homosexual relationship with a young man having connections with the Russian royal family.

✋ 7 November

ON this day . . . in 1917 the Communists came to power in Russia as Kerensky's government was overthrown in what is still known as 'the October Revolution', since according to the Julian calendar (still in use in Russia at that time) the date was 25 October; and Leon Trotsky (1879–1940), the People's Commissar for Foreign Affairs in the first Soviet government, was born on this day (according to the New Style calendar).

Also born on this day were: Marie Curie (1867–1934), the Polish-born French scientist who was the first woman to win a Nobel prize (in 1903, with her husband Pierre Curie and Antoine Becquerel) and the first double prizewinner (in 1911, for work on radium); Sir Chandrasekhara Venkata Raman (1888–1970), the first Indian scientist to win a Nobel prize (for physics, in 1930); the French novelist Albert Camus (1913–60), the recipient of the Nobel prize for literature in 1957; the American evangelist Billy Graham (1918–   ); and Dame Joan Sutherland (1926–   ), the Australian prima donna known as 'La Stupenda'.

✌ 8 November

ON this day . . . in 1923 Hitler's first attempt to seize power began with a meeting in a beer hall in Munich ('the Beer Hall Putsch'), which led to his imprisonment and the dictation of his political testament, *Mein Kampf*, to Rudolf Hess.

Born on this day were: Edmund Halley (1656–1742), the English astronomer who gave his name to the comet which reappeared sixteen years after his death; Sir Benjamin Hall (1802–67), the British politician who as the minister of works gave his name to the famous hour bell in the clock tower of the Houses of Parliament, Big Ben; Pierre Jean François Bosquet (1810–61), the French general remembered for his comment after witnessing the charge of the Light Brigade ('C'est magnifique, mais ce n'est pas la guerre.'); Herbert Austin, Lord Austin of Longbridge (1866–1941), the English car manufacturer who produced and gave his name to the popular 'Austin 7'; Sir Arnold Bax (1883–1953), the English composer and Master of the King's Music from 1942, who also wrote poetry using the pseudonym of 'Dermot O'Byrne'; the American author Margaret Mitchell (1900–49), whose novel *Gone with the Wind* (her only book) was immortalized through the film of the same name; and Christiaan Barnard (1922–   ), the South African surgeon who performed the first successful heart transplant operation (in 1967).

✌ 9 November

ON this day . . . in 1918 the German Kaiser Wilhelm was deposed and fled to Holland, two days before the signing of the Armistice; and in 1989 the newly elected East German government opened the borders with West Germany and the demolition of the Berlin Wall began.

Born on this day were: the Russian novelist Ivan Turgenev (1818–83); Edward VII (1841–1910), who succeeded his mother Queen Victoria in 1901; Sir Giles Gilbert Scott (1880–1960), the English architect who was the grandson of the architect Sir George Gilbert Scott; Jean Monnet (1888–1979), the French statesman and president of the European Coal and Steel Community (1952–5), known as 'the father of Europe'; and the American actress Katharine Hepburn (1909–   ).

Two British prime ministers, Ramsay MacDonald and Neville Chamberlain, died respectively in 1937 and 1940; and General de Gaulle, the first president of the Fifth Republic of France, died in 1970.

The Welsh poet Dylan Thomas died in America in 1953, aged 39 ('And death shall have no dominion.').

10 November

ON this day ... in 1885 Paul Daimler, the son of the German engineer Gottlieb Daimler, became the first motor-cyclist when he rode his father's new invention on a round trip of six miles.

Born on this day were: Martin Luther (1483–1546); the French composer François Couperin (1668–1733); George II (1683–1760), the last king of England to command his army in battle (at Dettingen, in 1743); the English painter and engraver William Hogarth (1697–1764); the Irish playwright and novelist Oliver Goldsmith (1728–74); the German playwright and poet Johann Christoph Friedrich von Schiller (1759–1805); the American novelist Winston Churchill (1871–1947); the American poet Vachel Lindsay (1879–1931); Sir Jacob Epstein (1880–1959), the American-born sculptor of Russo-Polish descent; Arnold Zweig (1887–1968), the German-Jewish novelist; the British actor Richard Burton (1925–84); and the British songwriter Tim Rice (1944–   ).

11 November

ON this day ... in 1918 World War I was effectively ended with the signing of the Armistice in the railway coach of the commander of the Allied armies, Marshal Foch, in the forest of Compiègne in France; and in 1920 the Cenotaph was unveiled by George V and the 'Unknown Soldier' was buried in Westminster Abbey.

Born on this day were: the French navigator Louis Antoine de Bougainville (1729–1811), the first Frenchman to circumnavigate the world; the Russian novelist Fyodor Mikhailovich Dostoevsky (1821–81); the French painter Jean Édouard Vuillard (1868–1940); the Swiss conductor Ernest Ansermet (1883–1969), founder of the Suisse-Romande

Orchestra; General George Smith ('Blood and Guts') Patton (1885–1945), the American soldier who commanded armoured corps in both World Wars; and Lord Jenkins (1920–   ), the British Labour minister who became the first leader of the SDP (the Social Democratic Party).

৯৯ 12 November

ON this day . . . in 1944 the *Tirpitz*, the last of the major German battleships, was sunk by bombers of the Royal Air Force.

Born on this day were: the English admiral Edward Vernon (1684–1757), who gave his nickname of 'Old Grog' to the diluted rum he issued to his sailors; Jacques Alexandre César Charles (1746–1823), the French scientist who invented and flew in the first hydrogen balloon; the Russian composer Alexander Borodin (1833–87); the French sculptor Auguste Rodin (1840–1917); Lord Rayleigh (1842–1919), the first Englishman to be awarded the Nobel prize for physics (in 1904); Ben Travers (1886–1980), the English dramatist who was awarded the CBE on the occasion of his 90th birthday (at which time he had three of his plays being performed in London theatres); and the American film actress Grace Kelly (1929–82), who became Princess Grace of Monaco from her marriage to Prince Rainier.

৯৯ 13 November

ON this day . . . in 1941 the British Navy's premier aircraft carrier HMS *Ark Royal* was torpedoed by a U-boat and sank off Gibraltar the following day.

Born on this day were: Edward III (1312–77), the father of Edward 'the Black Prince' and John of Gaunt; Sir John Moore (1761–1809), the posthumous hero of the battle of Corunna; John Peel (1776–1854), the eponymous hero of the famous hunting song by John Woodcock Graves ('D'ye ken John Peel with his coat so gray?'); James Clerk Maxwell (1831–79), the Scottish scientist who gave his name to the unit of magnetic flux; the Scottish novelist and poet Robert Louis Stevenson

(1850–94), who gave the English language the expression 'Jekyll and Hyde' from his novel *The Strange Case of Dr Jekyll and Mr Hyde*; Ludwig Koch (1881–1974), the German-born musician and naturalist who made the first out-of-doors recordings of wild birds; and George Carey (1935–   ), the 103rd archbishop of Canterbury.

✠ 14 *November*

Born on this day were: Leopold Mozart (1719–87), the Austrian composer and father and mentor of W. A. Mozart; Robert Fulton (1765–1815), the American engineer who invented the first successful steamship, the *Clermont*; John Curwen (1816–80), the English Congregational minister who invented the tonic sol-fa method of teaching sight-singing; Claude Monet (1840–1926), the French painter whose picture *Impression: Soleil levant* gave rise to the expression 'Impressionism'; the Belgian-born chemist Leo Hendrik Baekeland (1863–1944), whose invention of Bakelite gave rise to the modern plastics industry; Pandit Jawaharlal Nehru (1889–1964), India's first prime minister and the father of Mrs Indira Gandhi; Sir Frederick Grant Banting (1891–1941), the Canadian scientist who (with C. H. Best) discovered insulin; the American composer Aaron Copland (1900–90); Joseph McCarthy (1909–57), the US senator whose political witch-hunting gave us the word McCarthyism; Boutros Boutros-Ghali (1922–   ), the Egyptian secretary-general of the United Nations from 1992; King Hussein of Jordan (1935–   ); and HRH The Prince of Wales, Prince Charles (1948–   ), the heir apparent to the throne of Great Britain.

✠ 15 *November*

On this day . . . in 1968 the *Queen Elizabeth*, the largest passenger ship ever built, ended her last voyage as a passenger carrier.

Born on this day were: William Pitt 'the elder' (1708–78), the British prime minister 1756–7 and (as the Earl of Chatham) 1766–7; William Cowper (1731–1800), the English poet and hymn-writer; Sir William Herschel (1738–1822), the German-born astronomer who discovered the planet Uranus; the American poet Marianne Moore (1887–1972); Richmal Crompton (1890–1969), the English writer who created the perennial schoolboy 'William Brown' (who remained 11 years old from 1922 to 1969); the outstanding German general of World War II, Erwin Rommel (1891–1944), who was forced to commit suicide through his involvement in the attempted assassination of Adolf Hitler; Aneurin Bevan (1897–1960), the British Labour politician who introduced the National Health Service in 1948; the English poet and art critic Sir Sacheverell Sitwell (1897–1988), whose sister and brother were the writers Dame Edith and Sir Osbert Sitwell; and the Israeli pianist and conductor Daniel Barenboim (1942–   ), whose wife was the cellist Jacqueline du Pré.

❧ 16 November

ON this day . . . in 1632 Gustavus II of Sweden, 'the Lion of the North', was killed in defeating the Austrian General Wallenstein at the battle of Lützen.

Born on this day were: Tiberius (42 BC–AD 37), the second Roman emperor; Rodolphe Kreutzer (1766–1831), the French violinist and composer (and the dedicatee of Beethoven's Sonata for violin and piano, Opus 47); William John Thoms (1803–85), the English writer and bibliographer who originated the word 'folklore'; John Bright (1811–89), the English statesman and reformer and orator ('England is the mother of Parliaments'); William Frend De Morgan (1839–1917), the English artist and Pre-Raphaelite who became a successful novelist at the age of 65; William Christopher Handy (1873–1958), the American composer best remembered for the *St Louis Blues*; the American playwright George Simon Kaufman (1889–1961), best remembered for *The Man who Came to Dinner*; and the German composer Paul Hindemith (1895–1963).

Also born on this day were two British sportsmen: Willie Carson, the champion jockey (1942–   ); and the heavyweight boxing champion Frank Bruno (1961–   ).

✌ 17 November

ON this day . . . in 1558 Mary I died. ('When I am dead and opened, you shall find "Calais" lying in my heart.')

Born on this day was Field Marshal Bernard Law Montgomery, first Viscount Montgomery of Alamein (1887–1976), the commander in World War II of the British 8th Army, which drove General Rommel's forces from North Africa.

Also born on this day were: Louis Xavier Stanislas, Louis XVIII of France (1755–1824), the brother of Louis XVI and Charles X, known as 'Louis le Désiré'; and Lee Strasberg (1901–82), the American actor who evolved the training technique which became known as 'the Method'.

In 1941, the German World War I air ace Ernst Udet, a Luftwaffe general in World War II, was killed in a flying accident. (It has been said, however, that he took his own life to escape arrest by the Gestapo.)

✌ 18 November

ON this day . . . in 1477 William Caxton published the first dated book printed in England, a translation from the French of *The Dictes and Sayings of the Philosophers* by Earl Rivers.

Born on this day were: the Scottish artist Sir David Wilkie (1785–1841), the official painter to William IV (and whose burial at sea inspired the famous painting by Turner); Sir Henry Bishop (1786–1855), the composer of 'Home, Sweet Home' (featured in his opera *Clari*) and the first English musician to be knighted (in 1842); Louis Daguerre (1789–1851), the French painter who invented and gave his name to the first practical process of photography, the daguerrotype; John Nelson Darby (1800–82), the English theologian and founder of the Darbyites, an 'Exclusive' division of the Plymouth Brethren; Sir W(illiam) S(chwenck) Gilbert (1836–1911), the English playwright and librettist whose collaboration with the composer Sir Arthur Sullivan gave us the famous Savoy Operas; Ignacy Jan Paderewski (1860–1941), the Polish concert pianist and composer who became his country's first prime minister (in 1919); George Horace Gallup (1901–84), the American statistician who originated the public opinion poll; the American songwriter Johnny Mercer

(1909–76); and Alan B(artlett) Shepard (1923–   ), the first American to make a space flight.

℘ 19 November

On this day . . . in 1863 Abraham Lincoln delivered his address at the dedication of the National Cemetery at Gettysburg. ('. . . we here highly resolve that the dead shall not have died in vain, that this nation, under God, shall have a new birth of freedom; and that government of the people, by the people, and for the people, shall not perish from the earth.')

Born on this day were: James Abram Garfield (1831–81), the US president who was the second (after Lincoln) to be assassinated; and Mrs Indira Gandhi (1917–84), the Indian prime minister who was assassinated in her second term of office.

Also born on this day were: Charles I, the king of Great Britain and Ireland (1600–49); Viscomte Ferdinand de Lesseps (1805–94), the French engineer who conceived the plan for constructing the Suez Canal; and José Raoul Capablanca (1888–1942), the Cuban chess master and world champion 1921–7.

℘ 20 November

On this day . . . in 1759 the British fleet under Admiral Hawke defeated the French at the battle of Quiberon Bay and thwarted an invasion of England; and in 1917 tanks were deployed *en masse* for the first time as the battle of Cambrai began, with more than 400 British machines engaged in an offensive which breached the Hindenburg Line.

Born on this day were: Thomas Chatterton (1752–70), the English literary prodigy and forger; Edwin Hubble (1889–1953), the American scientist who gave his name to a law concerning the universe; Alistair Cooke (1908–   ), the English-born journalist, author, and broadcaster; and Robert Kennedy (1925–68), a younger brother of President Kennedy who was assassinated as he was campaigning for the Democratic candidacy.

✺ 21 *November*

On this day . . . in 1783 two Frenchmen, François de Rozier and the Marquis d'Arlandes, made the first free-flight balloon ascent in a hot-air balloon built by the Montgolfier brothers.

Born on this day were: the French writer and philosopher Voltaire, the assumed name of François-Marie Arouet (1694–1778); Victoria Adelaide Marie Louise, the Princess Royal (1840–1901), the first child of Queen Victoria and Prince Albert who became the mother of Kaiser Wilhelm II; Sir Leslie Ward (1851–1922), the caricaturist 'Spy'; Sir Arthur Quiller-Couch (1863–1944), the editor of *The Oxford Book of English Verse*; Sir Harold Nicolson (1886–1968), the diplomat and author (and husband of the novelist Victoria Sackville-West); René Magritte (1898–1967), the Belgian painter and caricaturist; Coleman Hawkins (1904–69), the American pioneer of the tenor saxophone in jazz music; and Malcolm Williamson (1931–   ), the Australian-born 'Master of the Queen's Music'.

✺ 22 *November*

On this day . . . in 1963 the US president John Fitzgerald Kennedy was assassinated in Dallas, Texas.

Born on this day were: the English composer and folk-song collector Cecil Sharp (1859–1924), whose parents named him from this day, St Cecilia's Day (St Cecilia being the patron saint of music), in lieu of an expected daughter; and the English composer Benjamin Britten, Baron Britten of Aldeburgh (1913–76).

Also born on this day were: Thomas Cook (1808–92), the English pioneer of the package tour; the English novelist Mary Ann Evans (1819–80), who adopted the pseudonym of George Eliot; Wiley Post (1889–1935), the American airman who was the first to fly solo around the world (in 1933); Charles de Gaulle (1890–1970), the French general who led the Free French forces from England in World War II and was the first president of the Fifth Republic (1958–69); Hoagy Carmichael (1899–1981), the American song composer best remembered for his song 'Stardust'; the Spanish composer Joaquín Rodrigo (1901–   ), best remembered for his

Guitar Concerto; the American tennis player Billie-Jean King (1943–), who won a record twenty Wimbledon titles (1961–79); Mushtaq Mohammad (1943–), the Pakistani cricketer who was the youngest Test player at 15 years and 124 days; and the German tennis player Boris Becker (1967–), who was the youngest Wimbledon champion at 17 years and 227 days.

In 1975 Prince Juan Carlos became King Juan Carlos I of Spain, two days after the death of General Franco; and in 1990 Margaret Thatcher, the first woman to become Prime Minister of Britain, announced she was to resign.

℀ 23 November

ON this day . . . in 1499 Perkin Warbeck, who had claimed to be Richard, Duke of York, the second son of Edward IV, was hanged at Tyburn after an unsuccessful attempt to escape from the Tower of London; and in 1910 'Doctor' Hawley Harvey Crippen, an American-born dentist, was hanged at Pentonville Prison for the murder of his wife, the actress and singer Belle Elmore, after an unsuccessful attempt to escape to Canada.

Born on this day were: Franklin Pierce (1804–69), the fourteenth US president; the American outlaw William H. Bonney (1859–81), who was known as 'Billy the Kid'; and the British film actor William Henry Pratt (1887–1964), who was known as 'Boris Karloff'.

Also born on this day was the Spanish composer Manuel de Falla (1876–1946); and the composer Thomas Tallis ('the father of English cathedral music') died in 1585. (His date of birth is not known.)

℀ 24 November

ON this day . . . in 1642 the Dutch navigator Abel Tasman discovered Tasmania, naming it Van Diemen's Land after the governor-general of the Dutch East Indies who had sponsored his expedition.

Born on this day were: the Dutch philosopher Baruch Spinoza (1632–77); the Irish-born novelist Laurence Sterne (1713–68), especially remembered for *The Life and Opinions of Tristram Shandy*; Zachary Taylor (1784–1850), the twelfth US president (known as 'Old Rough-and-Ready') and the second to die in office, the year after he became president; Grace Darling (1815–42), the daughter of a Scottish lighthouse keeper who became famous from her part in the rescue of survivors from a shipwreck in 1838; Frances Hodgson Burnett (1849–1924), the English novelist who became famous for *Little Lord Fauntleroy*; Scott Joplin (1868–1917), the American pianist and composer known as 'the king of Ragtime'; and the England cricketer Ian Botham (1955–   ).

In 1963 Lee Harvey Oswald, the alleged assassin of President Kennedy, was shot dead by a night club owner, Jack Ruby, as he was being taken to the county gaol.

✤ 25 November

On this day . . . in 1952 the first performance of Agatha Christie's play *The Mousetrap*, at London's Ambassadors Theatre, began the longest continuous run of any theatrical show in the world.

Born on this day were: the Spanish dramatist and poet Lope de Vega (1562–1635), founder of the Spanish national drama; the English actor and manager Charles Kemble (1775–1854); and the English actor-manager and playwright Harley Granville-Barker (1877–1946).

Also born on this day were: Andrew Carnegie (1835–1919), the son of a Scottish weaver who went to America and became a millionaire and one of the great philanthropists of his time; Karl Benz (1844–1929), the German motor manufacturing pioneer; Pope John XXIII (1881–1963); Isaac Rosenberg (1890–1918), the English poet and artist who was killed in action in World War I; and Joe DiMaggio (1914–   ), the American baseball player who hit in a record fifty-six games for New York in 1941 and later became the second husband of Marilyn Monroe.

In 1120, Henry I's only legitimate son, William was drowned when the 'White Ship' carrying him from Normandy to England sank off Barfleur. (His death caused the subsequent conflict for the English crown between Stephen and Henry's daughter Matilda.)

✌ 26 November

On this day . . . in 1789 the harvest of 1623 was first celebrated nationally in America. (The day of the week was Thursday and Thanksgiving Day has since been celebrated annually on the last Thursday in November.)

Born on this day were: William George Armstrong (1810–1900), later Baron Armstrong, the English inventor who produced the first hydraulic crane and a gun which was the prototype of modern artillery; Emlyn Williams (1905–87), the Welsh actor and playwright especially famed for his solo performances in the guise of Charles Dickens, Dylan Thomas, and Saki; and the Irish actor Cyril Cusack (1910–   ), whose three daughters are actresses and have all appeared with their father in a London production of Chekhov's *The Three Sisters*.

In 1703 Henry Winstanley, the English engineer who built the first Eddystone lighthouse, was among those who perished when it was destroyed in a gale, as he was supervising repairs to the structure.

✌ 27 November

On this day . . . in 1942 the French fleet was scuttled as the German forces entered Toulon.

Born on this day were: Anders Celsius (1701–44), the Swedish astronomer who invented and gave his name to the Celsius (or centigrade) thermometer; Mary Robinson (1758–1800), the English actress and writer (and mistress of the Prince Regent) known as 'Perdita', from the part she played in *The Winter's Tale*, who was painted by Reynolds, Romney, Gainsborough, and other noted artists of the time; the English actress and writer Fanny Kemble (1809–93), daughter of the actor Charles Kemble; Chaim Weizmann (1874–1952), the first president of Israel (1949–52); and Sir William Orpen (1878–1931), the Irish painter who was one of the first official war artists in World War I.

✌ 28 November

ON this day . . . in 1893 the women of New Zealand went to the polls in a general election (the first to do so); in 1919 the first woman to take her seat in the House of Commons, Nancy, Viscountess Astor, was elected as the Coalition Unionist member for the Sutton division of Plymouth in a by-election caused by the succession of her husband to the peerage; and in 1990 Mrs Margaret Thatcher, the first woman to become prime minister of Britain, delivered her resignation to the Queen.

Born on this day were: Jean Baptiste Lully (1632–87), the Italian-born French composer who founded the French national opera; William Blake (1757–1827), the English poet and artist; Sir Leslie Stephen (1832–1904), the first editor of *The Dictionary of National Biography* who was married to W. M. Thackeray's younger daughter and whose own younger daughter was the author Virginia Woolf; John Wesley Hyatt (1837–1920), the American inventor of the composition billiard ball; José Iturbi (1895–1980), the Spanish concert pianist who appeared in a number of Hollywood musicals; the English novelist Nancy Mitford (1904–73), who popularized the terms 'U' and 'non-U'; and the Italian novelist Alberto Moravia (1907–90).

In 1916 four people were injured in the first aeroplane raid on London.

✌ 29 November

ON this day . . . in 1945 Yugoslavia was proclaimed a republic.

Born on this day were: the Italian composer Gaetano Donizetti (1797–1848), best remembered for his opera *Lucia di Lammermoor*; Louisa May Alcott (1832–88), the American author best remembered for her novel *Little Women*; the English playwright Sir Francis Cowley Burnand (1836–1917), who wrote an operetta, *Cox and Box*, with Arthur Sullivan before the famous partnership of W. S. Gilbert and Sullivan came about; Sir John Ambrose Fleming (1849–1945), the British scientist who revolutionized radio telegraphy with his invention of the thermionic valve (in 1904); C(live) S(taples) Lewis (1898–1963), the British author and academic who is perhaps best remembered for books for children;

and Busby Berkeley (1895–1976), the American film director who revolutionized the song and dance routine into the Hollywood spectacle.

In 1530 Cardinal Thomas Wolsey died on his way to London to face a charge of high treason.

℀ 30 November

ON this day . . . in 1872 the first international football match was played, with Scotland (on St Andrew's Day) opposing England in Glasgow. (The match was drawn with no goals scored.)

Born on this day were: Sir Philip Sidney (1554–86), the poet and soldier; Jonathan Swift (1667–1745), the Irish writer best remembered for his satire *Gulliver's Travels*, for which he was paid £200 (the only payment he received for his writing); Frederick Temple (1821–1902), the ninety-fifth archbishop of Canterbury whose son William was to become the ninety-eighth archbishop; Samuel Langhorne Clemens, alias Mark Twain (1835–1910), the American writer and creator of 'Tom Sawyer' and 'Huckleberry Finn'; Angela Brazil (1868–1947), the English writer of school stories for girls; and L(ucy) M(aud) Montgomery (1874–1942), the Canadian writer whose *Anne of Green Gables* remains one of the best-known stories for girls.

In 1954 the British prime minister Sir Winston Churchill, on his 80th birthday, was presented with a portrait of himself by Graham Sutherland from both Houses of Parliament. (On the death of his wife the Baroness Spencer-Churchill, in 1977, it was revealed that the painting had been destroyed at Sir Winston's request.)

DECEMBER ❧

❧ 1 December

ON this day . . . in 1135 Stephen of Blois claimed the throne of England on the death of his uncle Henry I, repudiating the succession of Henry's daughter Matilda.

In 1581 Edmund Campion and other Jesuit martyrs were hanged at Tyburn for sedition, after being tortured.

Born on this day were: Madame Tussaud (1761–1850), the Swiss-born modeller in wax who founded the world-famous exhibition in London's Baker Street; Queen Alexandra (1844–1925), consort of Edward VII and mother of George V; the English author Henry Williamson (1895–1977), best remembered for *Tarka the Otter*; and Woody Allen (1935–), the American film actor, writer, and director.

In 1919 Lady Astor, the American-born wife of the second Viscount Astor, became the first woman to take her seat in the House of Commons; and in 1783 the French scientist J. A. C. Charles made the first ascent in a hydrogen balloon (of his own invention), ten days after the first manned ascent in a hot-air balloon.

❧ 2 December

ON this day . . . in 1805 Napoleon celebrated the first anniversary of his coronation with a decisive victory against the combined armies of Russia and Austria at the battle of Austerlitz (near what is now Brno in Czechoslovakia), also known as 'the battle of the three emperors'.

In 1859 John Brown, the American anti-slavery activist was hanged for treason at Charlestown, Virginia. ('But his soul goes marching on,' in the words of the famous Unionist song of the Civil War.)

Born on this day were: Dr Joseph Bell (1837–1911), the Edinburgh physician believed to have been the prototype of Sir Arthur Conan Doyle's detective 'Sherlock Holmes'; the French painter Georges Seurat (1859–91); Ruth Draper (1884–1956), the American character actress and monologuist; General Zhukov (1896–1974), the Russian commander of the final assault on Berlin in 1945; Harriet Cohen (1895–1967), the English concert pianist who was the dedicatee of Sir Arnold Bax's Concerto for Left Hand following an accident to her right hand; Sir John Barbirolli (1899–1970), the English conductor; Peter Carl Goldmark (1906–77), the Hungarian-born inventor of colour television and the long-playing record; and the American-born Greek opera singer Maria Callas (1923–77).

✌ 3 *December*

ON this day . . . in 1967 the first human heart-transplant operation was performed in Cape Town by Dr Christiaan Barnard, when the heart of a 24-year-old woman (killed in a road accident) was transplanted into a 55-year-old man. (The man died eighteen days afterwards.)

Born on this day were: Niccolò Amati (1596–1684), the most famous member of the Italian family of violin makers whose pupils included Stradivari and Guarneri; Samuel Crompton (1753–1827), the English inventor of the cotton-spinning machine named after him as Crompton's mule (since like the animal it was a hybrid of two other types); Sir Rowland Hill (1795–1879), the originator of the 'penny post'; Thomas Beecham (1820–1907), the English manufacturer and inventor of Beecham's pills, who was the grandfather of the conductor Sir Thomas Beecham; Joseph Conrad (1857–1924), the Polish-born naturalized British novelist; and the Austrian composer Anton Webern (1883–1945).

In 1800 the Austrians were defeated by the French under General Moreau at the decisive battle of Hohenlinden, near Munich.

✌ 4 *December*

ON this day . . . in 1791 the *Observer*, Britain's oldest Sunday newspaper, was first published.

Born on this day were: the Scottish historian Thomas Carlyle (1795–1881); Samuel Butler (1835–1902), the English writer who is known from the title of his satirical novel *Erewhon* (an anagram of 'nowhere'), to distinguish him from an earlier writer of the same name; Edith Cavell (1865–1915), the English nurse who was executed by the Germans for assisting the escape of Allied soldiers in Belgium; Wassily Kandinsky (1866–1944), the Russian painter and co-founder of 'Der Blaue Reiter' (The Blue Rider) group of artists; the German poet Rainer Maria Rilke (1875–1926); General Francisco Franco (1892–1975), the Spanish Fascist dictator; A(lfred) L(eslie) Rowse (1903–   ), the English historian and Shakespearian scholar; and the Canadian-born singer and film actress Deanna Durbin (1921–   ).

John Gay, the poet and playwright who wrote *The Beggar's Opera*, died in 1732. (He was buried in Westminster Abbey, where his monument bears his own epitaph: 'Life is a jest, and all things show it: I I thought so once, and now I know it.')

৯৯ 5 *December*

On this day . . . in 1872 the *Mary Celeste*, a ship which had left New York the month before, with the captain, his wife and daughter, and a crew of seven, was found abandoned off the Azores.

In 1791 Wolfgang Amadeus Mozart died in Vienna (aged 35) and was buried there in an unmarked pauper's grave.

Born on this day were: Martin Van Buren (1782–1862), the US president known as 'the red fox of Kinderhook' (his birthplace); Christina Georgina Rossetti (1830–94), the English poet and sister of the poet Dante Gabriel Rossetti; George Armstrong Custer (1839–76), the American soldier who was killed with all his men at the battle of Little Big Horn; Lord Jellicoe (1859–1935), the commander of the Grand Fleet at the battle of Jutland; Walt(er Elias) Disney (1901–66), the American producer of animated film cartoons and the creator of 'Mickey Mouse', 'Donald Duck', etc.; and the Spanish tenor José Carreras (1946–   ).

In 1933 prohibition in the USA was officially repealed after more than thirteen years.

✧ 6 December

ON this day . . . in 1877 the American inventor Thomas Alva Edison made the first known sound-recording when he recited 'Mary had a little lamb' into a phonograph of his own design.

Sir John Brown (1816–96), the English inventor who manufactured the first steel rails and originated armour plating for warships (in 1860), was born on this day.

Also born on this day were: Henry VI (1421–71), the youngest king of England to accede to the throne (aged 269 days); George Monck, first Duke of Albemarle (1608–70), the English soldier who fought with Charles I in the Civil War but changed sides after his defeat at Nantwich and was later the leading negotiator in the Restoration of Charles II; the French scientist Joseph Louis Gay-Lussac (1778–1850); Sir Osbert Sitwell (1892–1969), the English poet, playwright, and novelist and brother of Edith Sitwell; and Ira Gershwin (1896–1983), the American songwriter and brother of George Gershwin.

✧ 7 December

ON this day . . . in 1941 America was brought into World War II with the Japanese airborne attack on Pearl Harbor, in the Hawaiian Islands.

Born on this day were: Giovanni Lorenzo Bernini (1598–1680), the Italian sculptor and architect; Theodor Schwann (1810–82), the German scientist who discovered pepsin; Pietro Mascagni (1863–1945), the Italian composer best remembered for his opera *Cavalleria rusticana*; Willa Sibert Cather (1876–1947), the American authoress and Pulitzer prizewinner for 1922; and the Irish-born author Joyce Cary (1888–1957).

In 1815 Marshal Ney, one of Napoleon's leading generals and commander of the Old Guard at the battle of Waterloo, was shot for treason.

❧ 8 December

ON this day . . . in 1660 the first Shakespearian actress to appear on the English stage (she is believed to have been one Mrs Norris) made her debut as 'Desdemona'.

Born on this day were: the Roman poet Quintus Horatius Flaccus, known as Horace (65–8 BC); the Norwegian poet, dramatist, and novelist Bjørnstjerne Bjørnson (1832–1910), who wrote the words of his country's national anthem; the Finnish composer Jean Sibelius (1865–1957); the Czech composer Bohuslav Martinu (1890–1959); and James Thurber (1894–1961), the American artist and writer.

In 1914 the German cruisers *Scharnhorst*, *Gneisenau*, *Nurnberg*, and *Leipzig*, were sunk by a British force under Rear-Admiral Sturdee in the battle of the Falkland Islands, with the German Admiral von Spee going down in his flagship the *Scharnhorst*.

In 1941 Great Britain, Australia, and the United States declared war on Japan.

❧ 9 December

ON this day . . . in 1914 the first purpose-built aircraft carrier, HMS *Ark Royal*, was commissioned by the Admiralty and shortly afterwards saw service in the Dardanelles.

Born on this day were: the English poet John Milton (1608–74), who had been blind for a number of years when he wrote his epic poem *Paradise Lost* (for a total payment of £10); Claude Louis Berthollet (1748–1822), the French pioneer in chemistry; the French composer Émile Waldteufel (1837–1915), best remembered for *The Skaters' Waltz*; Joel Chandler Harris (1848–1908), the American writer who created 'Uncle Remus', 'Brer Rabbit', etc.; Clarence Birdseye (1886–1956), the American inventor of the deep-freezing process named after him; Dame Elizabeth Schwarzkopf (1915–   ), the German-born soprano; and Dame Judi Dench (1934–   ), the English actress.

🎝 10 December

ON this day . . . in 1941 the British battleships HMS *Prince of Wales* and HMS *Repulse* were sunk by Japanese air attacks in the battle of Malaya.

Born on this day were: the Belgian-born composer César Franck (1822–90), whose two operas, *Hulda* and *Ghiselle*, were not performed until some five years after his death; the American poet Emily Dickinson (1830–86), whose highly original and influential poems (which numbered more than a thousand) were not published until some five years after her death; Melvil Dewey (1851–1931), the American librarian who originated and gave his name to the decimal classification system; E(rnest) H(oward) Shepard (1879–1976), the English artist and cartoonist especially remembered for his original illustrations of the books of A. A. Milne and Kenneth Grahame; Viscount Alexander of Tunis (1891–1969), the British soldier who took his title from his part in the Allied victories in North Africa; and the French composer Olivier Messiaen (1908–92).

🎝 11 December

ON this day . . . in 1688 James II was forced to abdicate after William of Orange had landed in England on 5 November at the invitation of several British statesmen; and in 1936 Edward VIII abdicated after a reign of less than a year. ('I have found it impossible to carry the heavy burden of responsibility and to discharge my duties as King as I would wish to do, without the help and support of the woman I love.')

Born on this day were: Giovanni de' Medici (1475–1521), who as Pope Leo X excommunicated Martin Luther and bestowed the title of 'Defender of the Faith' upon Henry VIII; Charles Wesley (1757–1834), the English composer and organist and elder son of the hymn-writer Charles Wesley; the French composer Hector Berlioz (1803–69); the French poet, novelist, and playwright Alfred de Musset (1810–57); Robert Koch (1843–1910), the German scientist who was awarded the 1905 Nobel prize for physiology and medicine for his work on tuberculosis; and Alexander Solzhenitsyn (1918–), the Russian author who was awarded the 1970 Nobel prize for literature.

✣ 12 December

On this day . . . in 1901 the Italian inventor Guglielmo Marconi received the first transatlantic radio transmission in St John's, Newfoundland, from Polhu, in Cornwall.

Born on this day were: Erasmus Darwin (1731–1802), the English scientist and poet and grandfather of Charles Darwin; the French novelist Gustave Flaubert (1821–80); the Norwegian painter Edvard Munch (1863–1944); the American film actor Edward G. Robinson (1893–1973); the American singer and film actor Frank Sinatra (1915–); and the English playwright John Osborne (1929–), whose play *Look Back in Anger* (in 1956) gave him the sobriquet of 'the angry young man' of the British theatre.

✣ 13 December

On this day . . . in 1914 the first Victoria Cross to be awarded to a submariner (Commander Norman Douglas Holbrook) was won when the submarine B-11 dived under five rows of mines in the Dardanelles and torpedoed the Turkish battleship *Messudieh*; and in 1939 the battle of the River Plate began as the German pocket battleship *Graf Spee* was engaged by the cruisers HMS *Exeter*, HMS *Ajax*, and HMS *Achilles*, off Montevideo.

Born on this day were: Sir William Hamilton (1730–1813), the diplomat and archaeologist and husband of Emma, Lady Hamilton, Lord Nelson's mistress; Heinrich Heine (1797–1856), the German poet; the German engineer Ernst Werner von Siemens (1816–92); Alvin Cullum York (1887–1964), the American soldier who (as Sergeant York) was described by General Pershing as 'the greatest civilian soldier of the War'; and the English painter John Piper (1903–92).

✌ 14 December

ON this day . . . in 1911 a Norwegian expedition led by Roald Amundsen was the first to reach the South Pole, thirty-four days ahead of Captain Scott's party.

In 1918 women voted for the first time in a British general election and the first woman MP was elected, Constance, Countess Markievicz, the Sinn Fein member for St Patrick's, Dublin. (She was unable to take her seat, however, through being detained in Holloway Prison for political activities.)

Born on this day were: the French astrologer Michel de Nostredame, known as Nostradamus (1503–66), who prophesied correctly the manner of the death of the French king Henri II; the Danish astronomer Tycho Brahe (1546–1601); the English painter Roger Fry (1866–1934), who coined the term 'Post-Impressionism'; John Christie (1882–1962), the English patron of music who founded the Glyndebourne Festival Opera; and George VI (1895–1952), whose great-grandfather Prince Albert died on this day in 1861.

✌ 15 December

IN 1899 the Boers defeated the British under General Buller at the battle of Colenso (and seven Victoria Crosses were won in the action, including one by the son of Field Marshal Roberts, VC, and another by Captain W. N. Congreve, whose son was in turn to win the award in 1916).

In 1916 the first battle of Verdun came to its end with a French offensive. (The battle had begun in February and more than 700,000 German and Allied soldiers died in the action.)

Born on this day were: the Roman Emperor Nero (AD 37–68); the English painter George Romney (1734–1802); the French engineer Gustave Eiffel (1832–1923), who designed and gave his name to the famous tower in Paris; the French scientist Antoine Henri Becquerel (1852–1908), who gave his name to rays emitted by a radioactive substance; Lazarus Ludwig Zamenhof (1859–1917), the Polish philologist who invented the universal language Esperanto; and Rudolf von Laban (1879–1958), the

German choreographer who invented and gave his name to a system of notation for movements in ballet, Labanotation.

✌ 16 December

ON this day . . . in 1775 a party of young Americans, disguised as Red Indians, threw chests of tea from ships in Boston Harbor as a protest against the taxation imposed by the British Parliament under George III. (The 'Boston Tea Party', as it came to be known, was one of the major events leading to the American War of Independence.)

Born on this day were: Catherine of Aragon (1485–1536), the first wife of Henry VIII (and the widow of his elder brother Arthur), who bore him five children, only one of whom, the Princess Mary (later Mary I) survived; Gebhard Leberecht von Blücher (1742–1819), the Prussian soldier remembered for his decisive role at the battle of Waterloo (and the boots named after him); the German composer Ludwig van Beethoven (1770–1827); the English novelist Jane Austen (1775–1817); Johann Wilhelm Ritter (1776–1810), the German scientist who discovered ultraviolet rays; the Hungarian composer Zoltán Kodály (1882–1967); Sir John Berry ('Jack') Hobbs (1882–1963), the England cricketer who scored a record 197 First-Class centuries; and Sir Noel Coward (1899–1973), the English actor, playwright, and songwriter.

✌ 17 December

ON this day . . . in 1903 the American aviator Orville Wright made the first recorded flight in a powered aircraft, the flight lasting some twelve seconds with a height of eight to twelve feet. (His brother Wilbur made another flight on the same day lasting almost one minute.) In 1935 the Douglas Dakota (the DC-3), one of the most widely used aircraft of its time, was first flown.

Born on this day were: Prince Rupert (1619–82), a nephew of Charles I and generalissimo of the royalist army; the Italian composer Domenico Cimarosa (1749–1801); Joseph Henry (1797–1878), the American scientist who gave his name to the unit of inductance; the American poet and abolitionist John Greenleaf Whittier (1807–92); the English poet and novelist Ford Madox Ford (1873–1939); Mackenzie King (1874–1950), the Canadian statesman who was three times prime minister and held the office longer than any other Commonwealth prime minister; Arthur Fiedler (1894–1979), the American conductor of the Boston Pops Orchestra from 1930 until the last year of his life; and the English-born bandleader and composer Ray Noble (1903–78).

℘ 18 December

ON this day . . . in 1912 the discovery of the skull of a man near Piltdown, in Sussex, believed to have been some 50,000 years old, was officially announced at a scientific gathering. ('Piltdown Man', or *Eoanthropus Dawsoni*, after amateur archaeologist Charles Dawson, was eventually exposed as an elaborately planned hoax, which deceived the 'experts' for more than forty years.)

Born on this day were: Charles Wesley (1707–88), the youngest brother of John Wesley and the writer of more than 6,500 hymns; Joseph Grimaldi (1779–1837), the English comic actor whose first name came to be used as a synonym for a clown, a 'joey'; William Frederick Yeames (1835–1918), the English historical painter especially remembered for *When did you last see your father?*; Sir Joseph John Thomson (1856–1940), the English scientist who discovered the electron and was awarded the 1906 Nobel prize for physics; the English poet Francis Thompson (1859–1907); the English composer Lionel Monckton (1861–1924); the American composer Edward MacDowell (1861–1908); Hector Hugh Munro (1870–1916), the English novelist and short-story writer known as 'Saki'; and the English dramatist Christopher Fry (1907–   ).

℘ 19 December

ON this day . . . in 1562 the French Wars of Religion between the Huguenots and the Catholics began with the battle of Dreux.

Born on this day were: Sir William Edward Parry (1790–1855), the English sailor and Arctic explorer; Albert Abraham Michelson (1852–1931), the first American scientist to be awarded the Nobel prize (for physics, in 1907); Grace Mildmay (1900–53), the English opera singer and wife of John Christie, the founder of the Glyndebourne Festival Opera; Sir Ralph Richardson (1902–83), the English actor; Leonid Brezhnev (1906–82), the Soviet leader from 1964 until his death; and the French singer Edith Piaf (1915–63).

ॐ 20 *December*

ON this day . . . in 1939 Captain Hans Langsdorff, the captain of the *Graf Spee*, which he had ordered to be scuttled four days after the battle of the River Plate, shot himself in his hotel room in Montevideo.

Born on this day were: the English music-hall artist George Galvin (1860–1904), known as Dan Leno; Yvonne Arnaud (1890–1958), the French-born concert pianist and actress who gave her name to the repertory theatre in Guildford, Surrey; and the Duke of Kent (1902–42), the younger brother of George VI, who was killed on active service in a flying accident.

In 1952 a second coelacanth (a species which until 1938 was thought to have become extinct 50,000,000 years ago) was caught off Anjouan Island and was later named *Malania-Anjouanae*, after the island and Dr Malan, the South African prime minister, who had personally arranged for the fish to be flown to South Africa.

ॐ 21 *December*

ON this day . . . in 1620 an exploratory party from the *Mayflower* landed on Plymouth Rock and the date has since been celebrated as 'Forefathers' Day'.

Born on this day were: Benjamin Disraeli, first Earl of Beaconsfield (1804–81), the British prime minister and novelist; Joseph Stalin (1879–1953), the Russian dictator who took his name from his pre-revolutionary activities as the 'Man of Steel'; Dame Rebecca West (1892–1983), the British novelist who took her name from a character in Ibsen's play *Rosmersholm*; Walter Hagen (1892–1969), the American golfer who was four times the winner of the British Open Championship; and the British novelist Anthony Powell (1905–).

❦ 22 December

ON this day . . . in 1938 the first coelacanth to be identified (the fish was believed to have become extinct 50,000,000 years ago) was caught in the Bay of Chalumna, off South Africa, and was later to be named *Latimeria-Chalumnae*, in honour of a Miss Courtenay-Latimer who kept the fish for its subsequent identification.

Born on this day were: John Crome (1768–1821), the English landscape painter known as 'Old Crome' who founded the 'Norwich' school of painting; Frank Billings Kellogg (1856–1937), the US secretary of state who negotiated (with the French foreign affairs minister Briand) the Briand–Kellogg peace pact and was awarded the 1929 Nobel peace prize; the Italian composer Giacomo Puccini (1858–1924); the American poet Edwin Arlington Robinson (1869–1935), who was three times a Pulitzer prizewinner; and the English actress Dame Peggy Ashcroft (1907–91).

❦ 23 December

ON this day . . . in 1948 seven Japanese war-leaders, including the former prime minister Tojo, were hanged for war crimes; and in 1953 the former Soviet secret police chief Lavrenti Beria and six of his associates were sentenced to death and shot.

Born on this day were: Baron Axel Frederic Cronstedt (1722–65), the Swedish scientist who discovered nickel; Sir Richard Arkwright (1732–92), the English inventor of the spinning frame named after him; Joseph Smith (1805–44), the American founder of the Mormon Church; and Samuel Smiles (1812–1904), the Scottish writer and moralist ('He who never made a mistake never made a discovery.').

✇ 24 December

On this day... in 1914 the first air raid was made on England when a bomb was dropped from a German aeroplane on the grounds of a rectory in Dover. (There were no casualties.)

Born on this day were: King John (1167–1216), the youngest son of Henry II; St Ignatius of Loyola (1491–1556), the Spanish founder of the Jesuits; George Crabbe (1754–1832), the English clergyman whose poem 'The Borough' inspired Benjamin Britten's opera *Peter Grimes*; the French dramatist Augustin Eugène Scribe (1791–1861), who wrote the libretti for some sixty operas, including Auber's *Fra Diavolo*; Henry Russell (1812–1900), the English singer who composed the song 'A Life on the Ocean Wave', the marching tune of the Royal Marines; James Prescott Joule (1818–89), the English scientist who gave his name to the unit of electrical energy (the work done in one second by one ampere flowing through one ohm); Matthew Arnold (1822–88), the English poet and the eldest son of Dr Arnold of Rugby; Emanuel Lasker (1868–1941), the German mathematician and the world chess champion 1894–1921; and Sir (Michael) Colin Cowdrey (1932–   ), the England cricket captain whose father deliberately gave him the initials 'MCC' (for the Marylebone Cricket Club).

✇ 25 December

On this day... in 1914 on the first Christmas Day in World War I, British and German troops observed an unofficial truce on various parts of the Western Front and exchanged gifts and played football together in 'no man's land'.

In 1941 Hong Kong was surrendered to the Japanese; and in 1991 Mikhail Gorbachev, the leader of the Soviet Union for almost seven years, tendered his resignation (and the USSR virtually passed into history).

Born on this day were: the English scientist Sir Isaac Newton (1642–1727); the English poet William Collins (1721–59); Dorothy Wordsworth (1771–1855), the diarist and sister of William Wordsworth; Mohammed Ali Jinnah (1876–1948), the first governor-general of Pakistan; the American film actor Humphrey Bogart (1900–57); and Mohammed Anwar El-Sadat (1918–81), the president of the Arab Republic of Egypt 1970–81 and joint recipient of the 1979 Nobel peace prize.

✺ 26 December

On this day . . . in 1806 the advance of Napoleon's army was checked by the Russians at the battle of Pultusk; and in 1943, the German battle cruiser *Scharnhorst* was sunk off North Cape by the battleship HMS *Duke of York* and other British warships protecting a convoy bound for Russia. (This marked the virtual end of the era of the battleship.)

Born on this day were: Thomas Gray (1716–71), the English poet especially remembered for his 'Elegy Written in a Country Churchyard' ('Fair Science frowned not on his humble birth, I And Melancholy marked him for her own.'); the American novelist Henry Miller (1891–1980), especially remembered for his *Tropic of Cancer*; and Mao Tse-tung (1893–1976), the Chinese statesman who gave his name to the Chinese form of Communism.

✺ 27 December

On this day . . . in 1979 President Hafizullah Amin of Afghanistan was ousted and murdered in a coup backed by the USSR, beginning a war which lasted more than ten years.

Born on this day were: the German astronomer Johannes Kepler (1571–1630), who gave his name to the three important laws of planetary motion; Sir George Cayley (1773–1857), the designer of the first successful man-carrying glider; the French scientist Louis Pasteur (1822–95), who gave his name to pasteurization; and the German-born film actress Marlene Dietrich (1901–92).

The poet and essayist Charles Lamb died in 1834, a few months after the death of his last close friend, Samuel Taylor Coleridge ('some they have died, and some they have left me, I And some are taken from me; all are departed; I All, all are gone, the old familiar faces.').

᭍ 28 *December*

ON this day ... in 1879 more than seventy people died when the Tay Bridge, the first bridge to be built over the estuary of the River Tay, in Scotland, was blown down as a train was passing over it. (The eccentric versifying of the Scottish poet William McGonagall assured us that this date 'will be remembered for a very long time'.)

Born on this day were: Woodrow Wilson (1856–1924), the US president who took America into World War I and advocated the League of Nations; the English painter Wilson Steer (1860–1942); Sir Arthur Stanley Eddington (1882–1944), the English astronomer; Earl Hines (1905–83), the American jazz pianist and composer; and the English actress Dame Maggie Smith (1934–   ).

᭍ 29 *December*

ON this day ... in 1170 St Thomas à Becket, the thirty-ninth archbishop of Canterbury, was murdered in Canterbury Cathedral by four knights of Henry II. ('What a parcel of fools and dastards have I nourished in my house, that not one of them will avenge me of this one upstart clerk.')

Born on this day were: Madame de Pompadour (1721–64), the influential mistress of Louis XV, who gave her name to a hairstyle; Charles Macintosh (1766–1843), the Scottish inventor who gave his name to a waterproof cloth, mackintosh; Charles Goodyear (1800–60), the American inventor of vulcanized rubber for tyres; Andrew Johnson (1808–75), the American president who succeeded Lincoln on his assassination in 1865; William Ewart Gladstone (1809–98), the English statesman who was four times prime minister and retired (from the premiership) at the age of 84; Alexander Parkes (1813–90), the English chemist who invented celluloid; the Spanish cellist Pablo Casals (1876–1973); and (coincidentally and on the very same day) another nonagerian musician, the English viola player Lionel Tertis (1876–1975).

❧ 30 December

On this day... in 1460, at the battle of Wakefield (one of the major battles of the Wars of the Roses), the Duke of York (the father of Edward IV and Richard III) was defeated and killed by the Lancastrians.

Born on this day were: the French composer and conductor André Messager (1853–1929); Rudyard Kipling (1865–1936), the first Englishman to be awarded the Nobel prize for literature (in 1907); L(eslie) P(oles) Hartley (1895–1972), the English novelist and short-story writer; the Russian composer Dmitri Kabalevsky (1904–87); and Sir Carol Reed (1906–76), the English film director.

❧ 31 December

On this day ... in 1805 the French Revolutionary calendar (also known as the Republican calendar), which had been in use since 1793, was last used officially; and in 1960, in Great Britain, the 'farthing', which had been in use since the thirteenth century, ceased to be legal tender.

Born on this day were: Charles Edward Stuart (1720–88), the grandson of James II known as 'the Young Pretender' (as the claimant of the English throne), or 'Bonnie Prince Charlie'; Charles Cornwallis (1738–1805), the British soldier whose surrender to Washington at Yorktown, in 1781, effectively ended the War of Independence; the American soldier George Gordon Meade (1815–72), who defeated Lee at the battle of Gettysburg; John Taliaferro Thompson (1860–1940), the American soldier who invented and gave his name to the tommy-gun; the French painter Henri Matisse (1869–1954); the American soldier George Catlett Marshall (1880–1959), who originated and gave his name to the European Recovery Programme (the Marshall Aid Plan) and was the recipient of the 1953 Nobel peace prize; and the British actors Sir Anthony Hopkins (1937–) and Ben Kingsley (1943–).

INDEX

Barnum, Phineas 5 July
Barrie, Sir James 9 May
Barrow, Clyde 23 May
Barrymore, Ethel 28 Apr.
Barrymore, John 28 Apr.
Barrymore, Lionel 28 Apr.
Bart, Lionel 1 Aug.
Barthou, Jean Louis 9 Oct.
Bartók, Béla 25 Mar.
Barton, Sir Edmund 18 Jan.
Baskerville, John 28 Jan.
Bassey, Shirley 8 Jan.
'Bastille Day' 14 July
Bates, H. E. 16 May
Baudelaire, Charles 9 Apr.
Baum, Lyman Frank 15 May
Baum, Vicki 24 Jan.
Bax, Sir Arnold 8 Nov.; 2 Dec.
Baylis, Lilian 9 May
Bazaine, Marshal 18 Aug.
Beachy Head, battle of 30 June
Beardsley, Aubrey 21 Aug.
Beaton, Sir Cecil 14 Jan.
Beatrix, queen of The Netherlands
    31 Jan.
Beatty, David, Lord Beatty 17,
    24 Jan.; 31 May
Beaumarchais, Pierre Augustin
    Caron de 24 Jan.
'Beau Nash' 18 Oct.
Bechet, Sidney 14 May
Becker, Boris 22 Nov.
Becket, St Thomas à 29 Dec.
Beckett, Samuel 13 Apr.
Becquerel, Antoine 7 Nov.; 15 Dec.
Beecham, Thomas 29 Apr.; 3 Dec.
Beecham, Sir Thomas 29 Apr.;
    3 Dec.
Beeching, Dr Richard, Lord
    Beeching 21 Apr.
Beer, Israel see Reuter, Baron von
Beerbohm, Sir Max 22 Jan.; 24 Aug.
Beethoven, Ludwig van 20 Feb.;
    26 Mar.; 16 Nov.; 16 Dec.
Beeton, Mrs Isabella Mary 14 Mar.

Begin, Menachem 16 Aug.
Behan, Brendan 9 Feb.
Behring, Emil von 15 Mar.
Beiderbecke, Bix 10 Mar.; 7 Aug.
Beith, John Hay see Hay, Ian
Bell, Alexander Graham 3, 7,
    10 Mar.; 2 Aug.
Bell, Dr Joseph 22 May; 2 Dec.
Bellingham, John 11, 18 May
Bellini, Vincenzo 3 Nov.
Belloc, Hilaire 27 July
Bellow, Saul 10 June
Benchley, Nathaniel 8 May
Benchley, Peter 8 May
Benchley, Robert 8 May
Bengal, Nawab of 20, 23 June
Ben Gurion, David 16 Oct.
Benjamin, Arthur 18 Sept.
Bennett, Alan 9 May
Bennett, Arnold 27 May
Bennett, Floyd 9 May
Bennett, Richard Rodney 29 Mar.
Bennett, Tony 3 Aug.
Bennett, William Sterndale 13 Apr.
Benny, Jack 14 Feb.
Benson, Arthur Christopher 24 Apr.
Bentham, Jeremy 15 Feb.
Bentley, Edmund Clerihew 10 July
Benz, Karl 25 Nov.
Beresford, General 16 May
Berg, Alban 9 Feb.; 17 Oct.
Bergman, Ingrid 29 Aug.
Bergson, Henri 18 Oct.
Beria, Lavrenti 23 Dec.
Berkeley, Busby 29 Nov.
Berkeley, George 12 Mar.
Berlin, Irving 11 May; 13 Aug.
Berliner, Emile 20 May
Berlin Wall 30 May; 12 Aug.; 9 Nov.
Berlioz, Hector 11 Dec.
Bernadette, St 11 Feb.
Bernhard, Prince, of The
    Netherlands 29 June
Bernini, Giovanni Lorenzo 7 Dec.
Bernstein, Leonard 25 Aug.

CHRONOLOGY OF WORLD EVENTS

Palaeolithic Man

All dates BP (before present)

c.5,500,000–3,000,000	Early *hominids* (Australopithecines) evolve in African woodland; bipedal; herbivorous; found in East and Southern Africa.

Lower Palaeolithic

c.2,100,000	*Homo habilis* evolving in Africa, shaping and using stones as tools. Omnivore, killing small game; also a scavenger.
c.1,800,000	*Homo erectus* evolving in Africa descended from *Homo habilis*.
c.1,500,000	*Homo erectus* (formerly Pithecanthropus) disperses from Africa; uses fabricated stone tools; omnivore, killing small animals, scavenging remains of large ones; camps by lakes and in river valleys.
c.1,500,000	Acheulian hand-axes and cleavers in East and Southern Africa.
c.1,000,000	*Homo erectus* in East and SE Asia.
c.700,000	Evidence in Europe.
c.500,000	Fire?
c.250,000	*Homo erectus presapiens* emerging; evidence through Europe; possible hut-dwellings.

Middle Palaeolithic

c.120,000	*Homo sapiens* (anatomically modern humans) evolving in East and Southern Africa. Hand-axes and cleavers disappear; emphasis on tools made from stone flakes, probably hafted.
c.80,000	*Homo (sapiens) neanderthalensis* as cave-dweller in Europe; flint scrapers for preparing furs; burial of dead; absorbed(?) into populations of *Homo sapiens* by c.30,000.
c.50,000	*Homo sapiens sapiens* (modern humans) spreading from Africa through Asia, China, Australia, America (?c.25,000). Variety of tools (knives, axes, scrapers, harpoons, needles, awls, etc.) and materials (wood, bone, antler, stone, reed, leather, flint, fur, etc.).

Upper Palaeolithic

c.35,000	Early round houses in South Russia
c.30,000	First cave-painting and carving SW France (Lascaux). Female figurines across Europe and Russia.
c.27,000	First painting on stone tablets in Southern Africa (Namibia).
c.20,000	Finger-drawings on clay walls, e.g. Koonalda Cave, Australia. Paintings on rock surfaces in Australia.
c.15,000	Tools made on elongate blades of stone in Europe, Western Asia, East and Southern Africa. High level of art. Personal adornment (beads, pendants).
c.12,000	Siberia first peopled as ice age ends. Earliest potters in Japan. Microlithic tools widely made throughout the world. Bows, spears, knives in use.
c.10,000	Climatic change stimulates new economies and techniques. Beginning of farming in several parts of the world.

Near East, Mediterranean, and Europe	Rest of the World
All dates BC	
9000 Sedentary societies emerging: *Natufian culture* in Syria and Palestine; collection of *wild cereals*, first domestication of dog.	*Jomon culture* on coastal sites of Japan for fishing and collecting.

	Near East, Mediterranean, and Europe	*Rest of the World*
8300	*Post-glacial warming* and spread of forests begins.	
8000	*Pre-pottery neolithic* societies in Syria and Palestine, with cultivated cereals and mud-brick villages (Jericho).	*Cattle-keeping groups*, making pottery, spread into the Sahara, then wetter than today.
	Rapid *ice-retreat* in northern Europe and spread of light forest: *mesolithic* societies, e.g. *Maglemose culture* in birch and pine forest, hunting elk and wild cattle; first evidence of dugout *canoes*.	
7000	Domestication of *goat, sheep, pig* at various places in Near East, followed by *cattle*: animals used mainly for meat, not milk or wool. *Linen textiles*.	Cultivation of *root crops* in New Guinea and South America.
	First use of *copper* for small ornaments, made by hammering and heating pure copper. *Obsidian* imported to mainland Greece from island of Melos.	
6000	First pottery in Near East; use of *smelting* (lead).	Tropical *millet* first cultivated in the southern Sahara; temperate millet in China. *Wheat* and *barley* introduced to Pakistan.
	Major site of *Çatal Hüyük* in central Turkey. Domestication of *cattle*.	
	Farming spreads to SE Europe; first *neolithic cultures* there. Oak forests spread to northern Europe; *deer* and *pig* hunted.	
	Cattle first used for traction in Near East (*plough, sledge*). First evidence of *irrigation*, in Iraq.	
5000	*Copper-smelting* in Turkey and Iran (*Chalcolithic*). *Woollen textiles*; use of animals for *milk*; domestication of *horse* and *donkey*.	Beginning of *maize* cultivation in Mexico; *rice* cultivation in China. Rice cultivation in India.
	Tree crops (olive, fig, vine) cultivated in the Levant.	
	Growth of population in lowland Mesopotamia; *date-palm* cultivated.	
	Farming spreads into central Europe.	
	Use of *skin boats* in Baltic; permanent coastal fishing and collecting sites (*Ertebølle culture*); adoption of simple pottery.	
4000	*Copper-casting* and alloying in Near East; development of simple copper metallurgy in SE Europe. *Gold-working* in Near East and Europe.	*Llama* domesticated in highland Peru, as pack animal; cotton cultivated in lowland Peru.
	First *urban civilization* develops in Sumer, with extensive *irrigation*. Trading colonies established in Syria; temple-building, craft workshops with extensive importation of raw materials such as metals and precious stones. First *wheeled vehicles* and *sailing-boats* on Euphrates and Nile; use of *writing* (cuneiform script); spread of *advanced farming* (plough, tree crops, wool) and *technology* (wheel, alloy metallurgy) to SE Europe.	*Pottery* comes into use in South America. *Jade* traded in China.
	Farming spreads to western and northern Europe; construction of monumental *tombs* in *megalithic* technique in Portugal, Brittany, British Isles, Scandinavia.	
	Use of *horse* leads to expansion of first *pastoralist communities* on steppes north of Black Sea; burial mounds covering pit-graves. *Plough* and *cart* widely adopted in Europe.	
3200	Unification of *Upper and Lower Egypt*; trading expeditions by donkey up into the Levant, where urban societies now exist. *Troy* an important trading centre in north Aegean; *Cycladic culture* in the Greek islands. Copper-working in Iberia (Los Millares).	

Near East, Mediterranean, and Europe	*Rest of the World*
3000 Egyptian *hieroglyphic script* develops; *pyramid* building begins. Royal tombs in Mesopotamia (e.g. Ur) demonstrate high level of craftsmanship in secular city-states. Spread of burial mounds and *corded-ware culture* in northern Europe replaces megalithic tradition, though ceremonial monuments (Avebury, first phase of Stonehenge) continue in British Isles. Stone-built temples in Malta.	Introduction of *dog* to Australia. *Copper* and *bronze* metallurgy in China and SE Asia; *silk* production.
2500 Extensive Egyptian maritime trade with *Byblos* (Lebanon); *Ebla* a major centre in Syria, in contact with both Byblos and Mesopotamia. Exploration of eastern Mediterranean maritime routes along southern coast of Turkey to *Crete*, using boats with sails. Beaker cultures bring innovations (copper-working, horses, drinking cups, woollen textiles) to Atlantic seaboard.	Permanent *villages* with temple mounds and ceremonial centres in Peru. First *towns* in China (*Longshan culture*) with trade and specialized production.
2300 Empire of *Agade* (Akkad) unites Mesopotamian city-states. Trade with *Indus valley civilization* in Pakistan. *Akkadian* becomes diplomatic language of the Near East.	Spread of pottery-making and *maize* cultivation in Middle and South America.
2200 Collapse of Old Kingdom in Egypt, and of Empire of Agade. *Ur* revives as political centre in southern Mesopotamia.	
2000 *Middle Kingdom* established in Egypt; revival of Levantine cities (Ugarit); beginnings of *Middle Minoan* (palatial) period in Crete. Revival of northern Mesopotamian centres (Assur and Mari); Assyrian *merchant colonies* established in Anatolia. *Hittite Old Kingdom* develops; Babylonian empire expands.	Emergence of *Shang civilization* in China.
1700 Egypt dominated by Asiatic rulers ('Hyksos'). Cretan palaces reconstructed after earthquake; expanded trade-links with mainland Greece: growth of *Mycenaean civilization*. *Hittite empire* expands; Assyria dominated by *Mitanni*. Tin-bronze now standard in Europe. Appearance of *chariot*.	
1500 *Kassite dynasty* in Babylon; *New Kingdom* in Egypt following expulsion of the Hyksos; expansion of Egyptian empire in the Levant. Rulers buried in Valley of the Kings. During brief Armarna period, c.1350, Akhenaton introduces *monotheism* and founds new capital. Increasing conflict with Hittites. *Canaanite cities* flourish in the Levant, evolving the use of the *alphabet*.	Metal-working (copper, gold) in Peru. Expansion of *Lapita culture* into western Polynesia.
1200 General recession and political collapse in many parts of eastern Mediterranean: end of Mycenaean and Hittite palace centres, decline of Egyptian power; appearance of 'Sea Peoples', probably as foreign mercenaries. Spread of *iron* metallurgy. Expansion of nomadic Aramaean tribes in Levant. Temporary expansion of Assyria, and capture of *Babylon*. Expansion of agriculture and bronze-working in temperate Europe, associated with expansion of *Urnfield cultures*. Links between Cyprus and Sardinia, where *Nuraghic culture* develops.	*Olmec civilization*, with temple mounds and massive stone sculptures, in Mexico.

Near East, Mediterranean, and Europe	Rest of the World	Culture/Technology
1000 Development of *spice route* to Arabia; growth of coastal trade in Levant under *Phoenicians*; colonization of Cyprus and exploration of central and western Mediterranean.	*Zhou* (Chou) *dynasty* in China. *Adena culture* with rich burials under large mounds in Ohio and Mississippi valleys. *Chavín civilization* in Andes. Early cities in Ganges valley.	Hebrew and Greek *alphabets* developing from Phoenician.
David king of Israel (*c.*1005–970); makes Jerusalem his capital.		
Solomon king of Israel (*c.*970–930); extends kingdom to Egypt and Euphrates.		
Temple of Jerusalem built.		Worship of *Dionysus* enters Greece from Thrace.
930 Israel divides into *Kingdom of Israel* in the north (*c.*930–721) and *Kingdom of Judah* in the south.		Early Hebrew texts (Psalms, Ecclesiastes).
Phoenician contacts with *Crete* and *Euboia*.		
***c.*920** *Nubians* conquer *Egypt*; *Shabaka* rules from Thebes.		Early version of great Hindu epic the *Mahabharata*.
Celts move west into Austria and Germany.	Farming villages on *Amazon* floodplain.	*Geometric style* pottery in Greece.
858 *Assyrian empire* reaches Mediterranean.		
814 Legendary date at which Phoenicians found *Carthage*.		
750 *Greek colonies* in southern Italy (*Cumae*) and Sicily.		*The Iliad* and *Odyssey* emerge from oral tradition.
Greek city-state culture throughout Aegean; *Lydia* pioneers *coinage*. Colonies spread through Mediterranean; Sicily *Magna Graecia*.		
Nomadic *Scythians* from southern Russia invade Asia Minor.		
753 Legendary date for *foundation of Rome*.		776 *First Olympiad*.
721 Assyrians under *Sargon II* (721–705) conquer Israel; under *Sennacherib* (704–681) empire expands and *conquers Egypt*. *Nineveh* the Assyrian capital.		*Taoism* founded in China, traditionally by Lao-tzu.
700 *Hallstatt* culture (Celtic iron-age warriors) in Austria moves west and down Rhône to Spain.		
Carthage expands through western Mediterranean, occupying *Sardinia* and *Ibiza*.	*Zhou* dynasty in China establishes *legal system*.	*Iron technology* enters India.

Near East, Mediterranean, and Europe	Rest of the World	Culture/Technology
First Celtic *hill-forts*.	*Magadha kingdom* on Ganges in India.	*Zarathustra* in Iran.
		Dionysiac festivals in Greece leading to *drama*.
640 Assyrians under Ashurbanipal conquer *Elamites*.	660 traditional date for *Jimmu*, first *Japanese emperor*.	*Library* established in Nineveh under Ashurbanipal (*c.*668–627).
612 Assyria defeated by *Medes* and *Babylonians*; sack of *Nineveh*. Babylon under *Chaldean* dynasty.	New Nubian kingdom of *Cush* established at *Meroë* on Upper Nile.	
600		*Doric order* appears in Greek architecture.
		Trireme (warship) evolves.
594 Ionian Greeks found *Massilia* (Marseilles).		*Hanging Gardens of Babylon* built.
Solon reforms *Athenian law.*		Jeremiah writing.
586 Chaldean *Nebuchadnezzar* conquers Jerusalem. Jews to Babylon in *Captivity*.	King *Vishtaspa* of Persia converted to *Zoroastrianism*.	*Thales* of Miletus developing *physical science* and *geometry*.
Greeks colonize Spain; import *Cornish tin*.		
561 *Pisistratus* ('benevolent tyrant') controls Athens 561–527.	550 *Cyrus II* defeats Medes; establishes Persian empire from Susa; captures Babylon (539).	Greek lyric poetry (*Sappho*).
546 Persian empire extends to Aegean. *Ionian Greek cities* captured.		*Pythagoras* teaching in southern Italy.
538 *Jews return* and rebuild Temple.		*Ionic order* appears in Greek architecture.
509 *Roman republic* proclaimed; Etruscan rule ends.	Persian empire reaches *India*.	*Siddhartha Gautama* (the Buddha) teaching.
508 *Cleisthenes* establishes *democratic constitution* in Athens.	Chinese bronze coinage in form of miniature tools (knives and spades).	Athenian pottery at its zenith.
		Iron-working in China.
500 *Etruscans* at the height of their power.		Emergence of *Greek drama*; theatres built.
499 Revolt of *Ionian Greek* cities against the Persians.	First inscriptions from *Monte Albán*, Mexico.	Greek *philosophical thought* emerging (*Heraclitus* at Ephesus).
490 Persian emperor *Darius* invades Greece; his army defeated at *Marathon*.		*Confucius* (d. 479) teaching in China.
480 *Xerxes*, his son, crosses Hellespont by bridge of boats; invades Greece with army and navy. Allied with Thebes, he wins land battle at *Thermopylae*, devastates Attica, but is defeated by Greeks in sea battle at *Salamis*. Xerxes retreats.		Aesop: *Fables*.

Near East, Mediterranean, and Europe	Rest of the World	Culture/Technology
472	Athens controls Aegean through *Delian League*.		Aeschylus: *The Persians*.
450	*Rome* extending power in *Latium* and against Etruscans. *Twelve Tables* (set of laws) drawn up.	*Persian empire* in decline.	
443	*Pericles* dominates Athenian democracy until 429. Athens rebuilt. *Parthenon* built, designed by Phidias.	Hanno the Carthaginian sails to *Senegal*.	Sophocles: *Antigone*. Herodotus: *History*.
	Celtic *La Tène* culture flourishes in Switzerland.	*Coinage* reaches India. Extensive trade links between Mediterranean and Asia.	*Phidias* leading Greek sculptor. *Solar calendar* in China.
431	*Peloponnesian War* begins; Athens against Sparta and her allies (431–404), with brief interlude (421–415) after *Peace of Nikias*.		Democritus: *theory of atom*. Euripides: *The Trojans*.
415	*Alcibiades* leads Athenian expedition to Sicily. Disastrous siege of *Syracuse*; many Athenians put to death.	*Nok culture* in West Africa (northern Nigeria), lasting until AD c.200. Iron metallurgy; clay figurines.	Aristophanes: *Comedies*.
405	Spartan victory at *Aegospotami*; Athens sues for peace (404).		Thucydides: *History of Peloponnesian War*.
390	Celts cross Brenner and *sack Rome*.	In China *Zhou* dynasty in decline.	399 *Socrates* condemned to death in Athens.
			387 *Plato* founds *Academy* in Athens.
371	*Thebes* led by *Epaminondas* defeats Sparta at *Leuctra* and briefly dominates Greece.		*Hippocrates* developing medicine on *Cos*.
	Rome dominates *Latium*, building roads and aqueducts.		
338	*Macedonia* defeats Thebes at battle of *Chaeronea*; controls all Greece under *Philip II* (359–336).		*Iron metallurgy* in Central Africa.
334	*Alexander* crosses Hellespont, defeats Darius III at *Granicus*, liberates Ionian cities; captures *Tyre* and *Egypt*; marches east.	*Alexander* master of Persian empire; invades *India*.	335 *Aristotle* founds *Lyceum* in Athens.
323	*Death of Alexander* in Babylon. His adoption of Persian life-style had been resented in Macedon. Empire disintegrates. His general *Antigonus* master of Macedonia.	c.321 *Chandragupta Maurya* (d. 296) establishes empire in India, overthrowing *Magadha kingdoms* and advancing west into lands occupied by Alexander; *Pataliputra* capital.	*Epicurean* and atomistic theory fashionable. *Hellenistic art* spreads throughout Asia. Menander: *Comedies*.
311	*Seleucid* power established in Babylon.		*Zhuangzi (Chuang-tzu)* in China writing on *Taoism*.
304	*Ptolemy* founds dynasty in Egypt.		

Near East, Mediterranean, and Europe	Rest of the World	Culture/Technology
300 Hellenistic kingdom of Attalids established at *Pergamum*.	Early Classic *Maya* culture developing in Guatemala.	292–280 *Colossus* of Rhodes. 284 *Library* founded (100,000 volumes) at Alexandria. *Euclid* teaching there.
280 *Pyrrhus* of Epirus campaigns in Italy; defeats Romans in several battles but is unable to exploit victory and suffers heavy losses. He returns to Greece. *Rome* continues to advance into southern Italy.		*Catapult* and *quinquereme* warship invented at Syracuse. Elephants first used in battle.
264 *First Punic War* begins, Rome against Carthage. Rome expands navy and wins control of Sicily (but not Syracuse), Sardinia, and Corsica.	269 *Asoka* (c.269–c.232), Mauryan emperor of India. Enthusiastic convert to Buddhism.	*Zoroastrianism* spreading in Persia. Greek and Oriental cultures fusing in Hellenistic period. Theocritus writes *Idylls* idealizing bucolic life on island of Cos.
241 *Carthage admits defeat.*	256 *Zhou dynasty* ends in China.	
237 *Hamilcar* of Carthage conquers SE Iberia, based on Gades.	Extensive trade between China and Hellenistic world.	*Latin literature* beginning to emerge.
218 *Second Punic War* begins. *Hannibal* crosses Alps from Spain and wins battle at *Lake Trasimene*; defeats Romans at *Cannae*, but fails to take city.	221 *Qin Shi Huangdi* (*Ch'in Huang-ti*) establishes dynasty in China. Conquers Zhou provinces and unites country politically.	214 *Archimedes'* inventions resist Romans in siege of *Syracuse*. Archimedes put to death 212.
214 Romans fight *First Macedonian War* (214–205).	c.210 *Great Wall of China* constructed.	
202 *Scipio* defeats Carthage at *Zama* and Second Punic War ends (201).	206 *Han dynasty* established in China.	Horse collar and harness in China.
198 *Second Macedonian War* (198–196). Rome wins battle of *Cynoscephalae* (197) against Philip V.		
192 *Seleucid* Antiochus III occupies Athens and Greece.	c.184 *Mauryan empire* ends in India.	*Plautus* (d. 184) and *Terence* (d. 159) writing *comedies* in Rome.
172 *Third Macedonian War* (172–168/7). Rome wins battle of *Pydna* against Perseus (168). Macedonia subjugated.		
168 Revolt of Maccabees in Palestine against Seleucids; *Judas Maccabaeus* establishes Jewish dynasty in Jerusalem.	*Parthian empire* (c.250–AD c.230) at its height, from Caspian Sea and Euphrates to the Indus.	
149 *Third Punic War* begins.		
146 *Carthage destroyed*. Rome dominates western Mediterranean.		*Buddhism* spreading throughout SE Asia.
143 *Macedonia* becomes Roman province.		

Near East, Mediterranean, and Europe	Rest of the World	Culture/Technology
133 Rome master of *Iberia*, occupies *Balearic Islands* (123). The *brothers Gracchi* attempt social and legal reforms in Rome 133–121.	*127–101 Han armies* from China conquer Central Asia. Drift west of *Asiatic* tribes.	Polybius: *Histories* (40 volumes of Roman history, 220–146).
112 Outbreak of war between Rome and *Jugurtha*, king of Numidia.	Roman envoys to Han China.	*Parchment* invented in Pergamum to replace papyrus.
107 *Gaius Marius* first elected *Consul at Rome* (he was to be consul seven times before he died in 86 BC).	*Teotihuacán* and *Monte Albán* developing in Mexico.	
105 *Jugurtha* captured by Marius' quaestor Sulla.		
c.100 Celtic *Belgae* first settle in SE Britain.		Water-mill first described in Greek writings; came from China.

Mediterranean and Europe	Rest of the World	Culture/Technology
91 *Italian Confederacy* of tribes on Adriatic and Apennines rebel against Rome; civil war.		
90 *Lex Julia* extends Roman citizenship to the Latin and some Italian cities.		
88 *Mithridates VI*, king of Pontus, invades Greece; defeated by *Sulla* at *Chaeronea* (85).		*Buddhism* spreading in China.
73 *Spartacus* leads revolt of 40,000 slaves; suppressed by *Crassus* and *Pompey* (71); 6,000 *crucified* along Appian Way.		
66 *Mithridates* finally defeated by Pompey.	*Classic civilization of Peru* emerging (pyramids, palaces, etc.).	*Cicero* (106–43) pleading and writing in Rome.
63 *Catiline's conspiracy* at Rome.	Romans under *Pompey* conquer Syria and Palestine; end of *Seleucid empire*.	
60 *First Triumvirate* in Rome: Crassus, Pompey, and Caesar co-ordinate their political activities.		
58 Caesar campaigns against *Gauls*.		
55, 54 Caesar invades *Britain*.	*Teotihuacán civilization* of Mexico flourishes until 8th century. City of some 200,000 with complex of streets and apartment blocks. *Pyramid of the Sun* 700 ft. long and 200 ft. high.	

Mediterranean and Europe	Rest of the World	Culture/Technology
53 *Crassus* killed in battle against *Parthians*.		
52 *Vercingetorix* leader of Transalpine Gauls; defeated by *Caesar* (51).		
49 Caesar crosses *Rubicon* and begins civil war.		
48 Caesar defeats *Pompey* in battle of *Pharsalus*. Pompey murdered at Rome. Caesar campaigns in Egypt, Africa, and Spain (48–45).		*Library of Alexandria* burned, but re-established.
45 Caesar *dictator of Rome*.		46 Caesar introduces *Julian calendar*.
44 Caesar assassinated.		
43 *Second Triumvirate* (Octavian, Antony, Lepidus).		42 Virgil begins to write *Bucolics*.
31 Battle of *Actium*. Antony and Cleopatra commit suicide (30).		
27 Octavian accepts title of *Augustus*.		*Pantheon* built in Rome.
		19 Virgil dies, leaving *Aeneid* unfinished.
		17 Herod rebuilds Temple at Jerusalem.
6? *Jesus of Nazareth* born.	*c.9 Western Han* dynasty ends in China.	Strabo: *Geography*.
AD 9 Germans under *Arminius* annihilate 3 Roman legions. Rome withdraws to Rhine.		Vitruvius: *treatise on architecture*.
14 Death of *Augustus*. Julio-Claudian emperors follow. Empress *Livia* plots. *Tiberius* emperor (14–37).	*c.25 Eastern Han* dynasty established in China.	Ovid: *Metamorphoses*. *c.30 Crucifixion of Jesus*, followed by foundation of Christianity.
37 *Caligula* emperor; assassinated AD 41.		
41 *Claudius* emperor (AD 41–54).		
43 *Britain occupied* under Claudius.		
54 *Nero* emperor (AD 54–68).		*c.51 St Paul* writes first letters.
61 *Boudicca's revolt* in Britain crushed by Suetonius.	*Kingdom of Axum* (Ethiopia) flourishes from port of Adulis.	Cult of *Mithras* in Roman army.
64 Great fire of Rome: *Christians* blamed and *martyred* in Rome (64–7), including *St Paul* and *St Peter*.	66 *Revolt of Jews* in Palestine.	*Hero* invents various machines in Alexandria.
69 *Vespasian* emperor (69–79); first of *Flavian* emperors; restores imperial economy.	*Dead Sea Scrolls* hidden at Qumran near the Dead Sea.	*c.65 First Gospel* (St Mark).

Mediterranean and Europe	Rest of the World	Culture/Technology
79 Eruption of *Vesuvius*. *Pompeii* and *Herculaneum* buried.	70 *Destruction* of *Temple* at *Jerusalem* under Titus, son of Vespasian. Jewish *Diaspora*.	c.75 *Colosseum* in Rome begun. *Paper*, *magnetic compass*, and *fireworks* invented in China. 90 Plutarch: *Lives*.
98 *Trajan* emperor (98–117); extends empire, defeating Dacians, Armenians, and Parthians. Empire at its fullest extent.	*Christianity* spreading.	Reform of *Buddhism* in India. *Iron-working* in Zambia.
100 *Pax Romana* throughout Europe, North Africa, and Middle East.		Juvenal (d. 130): *Satires*. Synod of Jamnia (in Palestine) fixes canon of *Old Testament* for Judaism.
117 *Hadrian* emperor (117–38).		c.117 Tacitus: *Annals*.
122 *Hadrian's Wall* built in Britain against the Picts.		c.125 Gaius Suetonius: *Lives of Caesars*.
132 *Bar-Cochba* leads revolt of Jews against the Romans.	Teotihuacán civilization in Mexico flourishing.	
138 *Antoninus Pius* emperor (138–61). Founds Antonine dynasty; streamlines imperial government.		Cult of *Mithras* continues to spread.
c.140 *Antonine Wall* built in Britain; abandoned c.163.		
161 *Marcus Aurelius* emperor (161–80); a Stoic philosopher, he campaigns on eastern and northern frontiers of empire.	*Parthian empire* weakening.	*Astronomy* developing in school of Alexandria. c.180 Marcus Aurelius: *Meditations*.
180 *Commodus* emperor (180–92); murdered for his wild extravagance and cruelty.	184 'Yellow Turban' revolt in China as Han dynasty declines.	*Galen* from Pergamum (d. 199) practising medicine in Rome.
193 *Septimius Severus* emperor (193–211); resumes *persecution of Christians*; long campaign in Britain against Picts; dies in York.		c.197 Tertullian: *Apology* (defends Christianity against Greek philosophic thought).
211 *Caracalla* emperor (211–17); his reign one of cruelty and extortion; murdered on campaign in Parthia.	220 Han dynasty ends in China. 220–63 Period of the 'Three Kingdoms' in China.	
212 *Edict of Caracalla* extends *Roman citizenship* to all freemen of empire.		*Manes* (c.215–75) from Persia develops *Manichaeism*. Egyptian *Coptic Church* to Ethiopia.
218 *Elagabalus* emperor (218–22); wild and decadent; seeks to impose worship of Syrian sun-god Elah-Gabal. Murdered in Rome.		*Neoplatonism* in Alexandria.

Mediterranean and Europe	Rest of the World	Culture/Technology
222	*Alexander Severus* emperor (222–35); under strong influence of Empress Mamaea; rule remembered as just; re-established authority of Rome.	224 *Ardashir* of Persia defeats Parthians, whose empire collapses. The Persian *Sassanid empire* established.	Indian art (sculpture and painting) flourish. Chinese literature developing.
235	Political tensions in Rome as empire begins to decline.	232 Sassanid Ardashir II defeated by Emperor Alexander Severus.	Gnostic Manichaeism spreads from Persia.
249	*Decius* emperor (249–51); intense persecution of Christians; Danube and the Balkans overrun by Goths.	c.250 Syrian kingdom of Palmyra rises in power under *Odaenathus* (d. 267).	Roman *architecture* covers Europe and Mediterranean.
253	*Valerian* (253–60) and *Gallienus* emperors (253–68). *Franks* invade empire and Sassanids take Syria. Gallienus defeats Alamanni (258).	259 Sassanid *Shapur I* captures Valerian. Bantu tribes move into Southern Africa.	*Christian theology* emerging in Asia Minor and Egypt.
270	Emperor *Aurelian* (270–5) abandons Dacia to Goths but *regains Rhine and Danube.*	273 Kingdom of *Palmyra conquered* by Roman emperor Aurelian.	
284	*Diocletian* emperor (284–305); re-establishes frontiers and reorganizes government.	Classic *Maya civilization* emerging (Tikal, Uaxactún, Palenque); first Maya stele from Tikal 292.	Monastic ideal (hermits) becoming popular (cf. *St Anthony*). c.285 *Pappus Alexandrinus* last great mathematician of Alexandria.
286	*Aurelius Carausius* seeks to separate Britain from empire.		
293	Diocletian establishes *Tetrarchy*; he rules with *Galerius* in east, *Maximian* and *Constantius* in west.		*Arius* of Alexandria (c.250–c.336) founds *Arianism*, denying divinity of Christ.
296	*Constantius* re-establishes control of *Britain.*		
303	Diocletian *persecutes Christians.*	Classic *Maya Old Empire* civilization (peaceful, civilian, highly literate; high levels of art; brick temples; priests very powerful with complex *religious ceremonies* and music).	
306	*Constantius* campaigns in Scotland. He dies. *Constantine* proclaimed in York.		
312	Constantine victorious at *Mulvian Bridge* against rival Maxentius.		
313	*Edict of Milan* allows liberty of cult of Christianity.		
315	*Constantine* (303–37) *sole emperor.*	317 Foundation of *Eastern Jin* (*Chin*) dynasty in China (317–420).	

Mediterranean and Europe	*Rest of the World*	*Culture/Technology*
324	Byzantium becomes capital of Empire as New Rome; named *Constantinopolis* (330).	320 Foundation of *Gupta empire* (to *c.*550) by *Chandra Gupta I* in India (golden age of religion, philosophy, literature, and architecture).	325 *Council of Nicaea* denounces Arianism and agrees *Creed*.
			Chinese mathematics reducing fractions and solving linear equations.
353	*Constantius II* sole emperor (353–60); re-establishes control over empire and defeats Alamanni at *Battle of Strasbourg* (357).		*c.*330–71 *St Martin* bishop of *Tours*.
	Saxons invading coasts of Britain.		Chinese *bucolic literature* flourishing.
360	*Julian (the Apostate)* emperor (360–3).		Julian (the Apostate) restores paganism briefly.
368	Order restored in Britain by *Count Theodosius*.		
374	*St Ambrose* elected bishop of Milan (d. 397); influences emperor and dominates western Church.		
378	*Visigoths* defeat Roman army at Adrianople.		
379	*Flavius Theodosius (the Great)* emperor (379–95). Pious Christian; defeats usurpers and makes treaty with Visigoths (382).	*Huns* from Asia concentrate on River Volga; moving west.	
391	All *pagan cults banned* in empire by Theodosius.		Greek and Latin Fathers continue to define Christian theology; *Athanasian Creed* agreed.
395	Empire divided on death of Theodosius: *Honorius* (395–423) rules from Milan, *Arcadius* (395–408) from Constantinople.		
396–8	Roman victories in Britain against Picts, Scots, and Saxons.		*c.*400 Text of Palestinian *Talmud* finalized.
404	Western capital moves from Milan to *Ravenna*.		*c.*405 *St Jerome* (d. 420) completes *Vulgate* (Latin Bible).
407	*Constantius III* proclaimed in Britain.		
410	*Visigoths*, led by Alaric, *sack Rome*. Romans *evacuate Britain*. *Franks* occupy northern Gaul and *Celts* move into Breton peninsula.	420 *End of Eastern Jin dynasty* in China.	
476	*Romulus Augustulus*, last western emperor (475–6), is *deposed*.		*Shinto* (worship of sun-goddess, reverence of ancestors and nature-spirits) in Japan.
	Saxon settlement in Sussex.		

Mediterranean and Europe	Rest of the World	Culture/Technology
481 *Clovis I*, king of Salian Franks (481–511), defeats Alamanni (496) and Visigoths (507).	Ecuadorean pottery dated *c.*500 found in *Galapagos Islands*. Evidence of *Pacific trade*?	*Buddhism* dominant in China.
488 Theodoric and *Ostrogoths* invade Italy.		
496 *Clovis baptized*; establishes *Merovingian* Frankish kingdom.		
Wessex occupied by Saxons.		
*c.*500 *British victory* over Saxons at Badon Hill.		
Visigoths established in Spain; Vandals in Africa (429–534).		

Britain and Europe	Rest of the World	Culture/Technology
552 *Ostrogoths* finally defeated by General *Narses*.	*c.*550 End of *Gupta empire* in India following attacks by Huns.	*c.*550 *St David* founds monastery.
560 *Aethelbert* king of Kent.		*Buddhism* in Japan along with *Shintoism*.
563 *St Columba* founds Iona.		
568 *Lombards* invade northern Italy.	*c.*570 Birth of *Muhammad*.	*Byzantine architecture* spreads throughout eastern empire and southern Italy.
Anglo-Saxon kingdoms in Britain emerging.		
Venice established as retreat from Lombards.		
577 *West Saxons* take Bath and Gloucester.	Chinese *Sui dynasty* 581–618; this short-lived dynasty reunites country and *rebuilds Great Wall*.	
590 Election of *Gregory the Great* as pope (590–604).	592 Prince *Shotoku Taishi* in Japan establishes mandarin-style bureaucracy.	*Gregorian chant* and Roman ritual, imposed by Gregory the Great.
	607 Unification of *Tibet*, which becomes *centre of Buddhism*.	
	618 *Tang dynasty* established in China (618–907); strong centralizing power restores order.	
627 Death of *Raedwald* of East Anglia. Possible burial at *Sutton Hoo*.	622 *Hegira* of Muhammad and friends: *Mecca* to *Medina*.	Christian missionaries to Germany and England (*St Augustine*, 597).
	632 *Death* of Muhammad.	
Saxon kingdoms in Britain (*Mercia*, *Wessex*, and *Northumbria*) struggle for power.	In Mexico Maya civilization at height; temples and palaces in stone (complex astronomical and mathematical knowledge).	*c.*625 Isidore of Seville: *Etymologies*.
		Parchment displacing papyrus.
629 *Dagobert*, king of Franks (629–39), reunites all Franks.		*c.*632 *Abu Bakr* collects the 114 chapters of the *Koran*.

255

Britain and Europe	Rest of the World	Culture/Technology
635 *Cynegils* of Wessex baptized.	634 *Rapid spread of Islam* in Arabia, Syria, Iran, N. Africa under *Caliph Umar* (634–44).	c.641 The great *Library of Alexandria* destroyed by Arabs.
		644 *Windmill* recorded in Persia.
	636 Collapse of *Sassanid* empire.	
664 *Synod of Whitby*; Roman practice imposed.	660 Damascus capital of *Ummayad empire of Islam*.	
669 *Archbishop Theodore* in Britain.		
678 Constantinople successfully *resists Arabs*.		680 Divisions within Islam produce *Sunnites* and *Shiites*.
679 *Battle of Trent*: Mercia becomes major British power.	*Teotihuacán civilization* of Mexico declining, possibly owing to destruction of rain forests.	
685 *Battle of Nechtansmere*: Picts defeat Northumbrians.	*Monte Albán*, Zapotec civilization, flourishing; influenced by *Teotihuacán*.	
Wessex expanding: Kent, Surrey, Sussex taken.		
687 *Pepin II* reunites Merovingian kingdom.	*Afghanistan* conquered by Arabs who cross Khyber Pass and conquer the *Punjab*.	*Grand mosques* of Jerusalem (692) and Damascus (706–15) built.
711 Muslim Arabs enter *Spain*, conquer Seville (712).	*Hsuan Tsung* (Tang emperor, 712–56) suffers drastic incursions by Arabs and revolts.	
718 Bulgars pressing south towards Constantinople.		
732 *Battle of Tours*; decisive victory of *Charles Martel*, 'Mayor of Palace' of Merovingians, over Muslims.	8th c. *Kingdom of Ghana* established; to last till 1240.	*Block printing* in China for Buddhist texts.
		731 Bede: *Ecclesiastical History*.
751 *Pepin III* (the Short), son of Charles Martel, ousts last Merovingian, *Childeric III*, and founds *Carolingian dynasty*.	Establishment of *Abbasid Caliphate* in Baghdad, 750–1258.	Golden age of Chinese poetry (*Li Po*, 701–61) and art (*Wu Tao-tzu*, d. 792).
Exarchate of Ravenna lost by Byzantines to *Lombards*. Pope turns to Franks for protection.		
754 *Pepin the Short* crowned in *St Denis* by Pope Stephen II; recognizes Papal States.		*Cordoba* centre of Muslim culture in Spain.
756 Abd al-Rahman founds *Caliphate of Cordoba*.		Irish *Book of Kells*.
757 *Offa* king of Mercia (757–96).		
768 *Charlemagne king of Franks* (768–814); campaigns against Avars and Saxons in east.		*Offa's Dyke* built.
		Dravidian temples in India.

256

Britain and Europe	Rest of the World	Culture/Technology
774 Charlemagne annexes Lombardy, but checked in Spain (death of *Roland* at *Roncesvalles* 778).	Caliph *Haroun al-Raschid* (786–809) establishes close links with Constantinople and with Charlemagne. Patronage of learning and the arts (*1001 Nights*).	In Spain *cotton* grown.
794 *Viking raids* on England and Ireland; Jarrow and Iona (795) sacked.	Kyoto capital of Japan, dominated by *Fujiwara* family.	
800 *Charlemagne crowned* in Rome by *Pope Leo III*.		*Alcuin* at court of Charlemagne. 'Carolingian Renaissance'.
		805 *Aachen cathedral* inspired by Byzantine models.
812 *Treaty of Aix-la-Chapelle* (Aachen). Charlemagne recognized emperor of the West by eastern emperor Michael I.		*Beowulf*, Anglo-Saxon poem.
813 Byzantine army defeated by Bulgars at *Adrianople*. Constantinople besieged by Bulgars and Arab army.		
825 *Wessex* annexes *Essex*.		
827 Byzantine loss of *Sicily* and *Crete* to Arab Saracens.		833 *Observatory* in Baghdad; Arabs develop astronomy, mathematics (*algebra* from India), optics, medicine.
		Romanesque architecture developing in West.
843 *Treaty of Verdun*; Carolingian empire divided: *East Franks, West Franks*, and *Lotharingia*. *Vikings* trading to Volga and Baghdad.	849 *Burma* unified by *Burmans* from Pagan.	843 Restoration of cult of *images* in Byzantine Church. *Icon art* to influence West through Venice.
		845 *Buddhism outlawed* in China.
		c.850 *Windmills* in Europe.
862 *Novgorod* founded as trading centre; Viking and Byzantine merchants.	857 *Fujiwara family* in Japan extend power.	c.850 China develops *gunpowder*.
St Cyril's mission to Moravia; Bulgars accept Christianity.		
866 Danish *Great Army* lands on east coast of Britain; Northumbria conquered; *Danelaw* established.	*Tiahuanaco* in Andes flourishing city.	
867 *Basil I* founds Macedonian dynasty of Byzantine empire (867–1057).		
869 *St Edmund murdered* by Danes.		
871 *Alfred the Great* king of Wessex (871–99).		

	Britain and Europe	Rest of the World	Culture/Technology
878	Alfred wins battle of *Edington*; Guthrum baptized.		
*c.*880	*Kingdom of Kiev* established in Russia.		
885	Paris *besieged* by Vikings.		
896	*Magyars* settle in *Hungary*.	889 Classic Old Maya civilization in Mexico *ending*.	
		*c.*900 Teotihuacán civilization ends; rise of *Toltecs* in North Mexico based on Tula.	
910	Danelaw conquered by *Edward* and *Ethelred*.	907 Tang dynasty in China ends; *China fragments*.	Monastery of *Cluny* established. *Benedictine Order* spreads through Europe.
		908 Shiite dynasty from Morocco, the *Fatimites*, conquer North Africa.	
911	Treaty of *St-Clair-sur-Epte* establishes *Rollo the Norseman* in NW France; Duchy of *Normandy*.		
919	Vikings establish *Kingdom of York* under *Ragnald*.		
926	*Athelstan* king of England (926–39).	Arabs trading along *East African* coast.	*Stone* replacing *wood* as building material in western Europe.
936	*Otto I* king of East Franks and Saxons (936–73).	935 *Koryo kingdom* in Korea (935–1392), capital Kaesong; Buddhism flourishes.	Expansion of European agriculture.
937	Athelstan defeats Vikings and Scots at *Brunanburh*.		
939	*Edmund* king of England (939–46).		
945	Norsemen in *Constantinople* and *Kiev*.	947 *Liao dynasty* from *Manchuria* (947–1125) extends control over North China.	943 Dunstan (*c.*909–88) abbot of Glastonbury. Under his leadership *monasticism* re-established in England.
954	Viking kingdom of York *ends*.		
955	Decisive battle of *Lechfeld*. Otto I defeats *Magyars*.		
959	*Edgar* king of England (959–75).		
960	*Dunstan archbishop* of Canterbury.	960 *Sung dynasty* in China till 1279; gradually reunites country; high levels of art and literature.	
962	Otto I crowned *Holy Roman Emperor* in Rome by Pope John XII. Seeks to establish power in Italy.		
971	*Bulgaria and Phoenicia* conquered by Byzantine armies.	969 *Fatimites* conquer W. Arabia, Syria, and Egypt; *Cairo* capital (973).	*c.*970 Synod of Winchester approves Ethelwold's *Regularis Concordia*.
973	Dunstan *crowns* and *consecrates* Edgar.		

Britain and Europe	Rest of the World	Culture/Technology
975	*Edward the Martyr* king (975–8); murdered by half-brother Ethelred.		
978	*Ethelred (the Unready)* king of England (978–1016).		
987	*Hugo Capet* king of France (987–96).	986 Viking settlements in Greenland. *New Maya Empire* emerging under Toltec influence.	988 Baptism of *Vladimir*, prince of Kiev (956–1015).
991	English treaty with Normans.	990 Ghana conquers Berber *kingdom of Audaghos* and gains gold and salt monopoly.	990 *School of Chartres* founded: early centre of western learning.
996	*Otto III* crowned in Rome (996–1002); establishes *capital in Rome* (999); with Pope Sylvester II aims to create universal Christian empire. *Balearic Islands* conquered by Caliphate of Cordoba.		*Avicenna* (980–1037) has lasting influence on West: philosophy, *Ash-Shifa* (Neoplatonic), and medicine *Qanun*; via Cordoba.
1001	Christian *kingdom of Hungary* established.		c.1000 Arabic description of magnifying properties of *glass lens*.
1003	*Swein Forkbeard*, king of Denmark, attacks England.	c.1000 *Chola Tamil dynasty* in S. India peaks under *Rajaraja I* (985–1016); stretches from Ganges to Malay archipelago.	
1013	*Swein invades England* and takes London.		
1014	Death of Swein. His son *Cnut* elected king by Danes in England.		
1016	Death of Ethelred; his son *Edmund Ironside* killed in battle. *Cnut* accepted as king of England (1017–35). Byzantine empire at height of power and influence under *Basil II* (976–1025).	*Mahmud of Ghazni* (971–1030) extends *Ghaznavid* empire into Persia and Punjab from Afghanistan.	
1035	*Harold* king of England (1035–40).		
1040	*Harthacnut* king of England and Denmark (1040–2).		
1042	*Edward the Confessor* king of England (1042–66).		c.1045 Printing by *movable type* in China.
1050	Bohemia, Poland, and Hungary become *fiefs* of Holy Roman Empire.	*Toltecs* flourish in Mexico; conspicuous Maya influence.	c.1050 *Salerno medical school* emerging; Arabic expertise. Cult of *Quetzalcóatl* in Toltec Mexico.
1054	*Schism* within Christian Church. Orthodox Eastern churches split from Catholic Rome.		

Britain and Europe	Rest of the World	Culture/Technology
1066 Normans conquer England. William I king (1066–87).	1064 Seljuk Turks menace Byzantine empire.	1063 St Mark's, Venice, rebuilt. 1063 Pisa Cathedral.
1069 'Harrowing of the North' by William I.	1068 Almoravid Berber dynasty in N. Africa; fanatical Muslims; build Marrakesh.	
1071 Normans established in southern Italy.	Battle of Manzikert; Seljuk Turks rout Byzantine army and threaten Asia Minor.	
1073 Hildebrand elected Pope Gregory VII (1073–85). 'Investiture Controversy' with Emperor Henry IV (1056–1106).		
1077 Henry IV at Canossa accepts papal supremacy to invest bishops of Church.		Bayeux Tapestry.
1081 Normans invade Balkans; Venice aids Constantinople against Normans; Venetian trade expands.	1080 Under Malik Shah Turks control Asia Minor and interrupt Christian pilgrim routes to Jerusalem.	c.1080 Chanson de Roland.
1084 Foundation of Grande Chartreuse (Carthusian Order of monks).		
1086 Domesday survey in England.		
1087 William II king of England (1087–1100).		Omar Khayyám (c.1050–1123): algebra; astronomy; poetry (Rubáiyát). 1090 Water-powered mechanical clock in China.
1095 Pope Urban II preaches Crusade at Clermont to rescue Holy Places from Turks.		1094 St Anselm: Cur Deus Homo?
1098 Cistercian Order of monks founded. Fairs of Champagne; urban society developing in Flanders, Germany, and Italy.	1099 Jerusalem captured by Crusaders.	Feudal system well established throughout Europe.
1100 Henry I king of England (1100–35).	Christian States in Palestine.	
1106 Henry V German emperor (1106–25).		
1108 Louis the Fat king of France (1108–37); Capet power expanding.		1115 St Bernard (1090–1153) abbot of Clairvaux; stresses spiritualism of monasticism.
1120 White Ship disaster; Henry I's son William drowned.	1116 Jin (Chin) dynasty in Manchuria. 1118 Knights Templar founded in Jerusalem.	'Twelfth-Century Renaissance'; rediscovery of Aristotle.

	Britain and Europe	Rest of the World	Culture/Technology
1122	*Concordat of Worms*; Henry V confirms end of Investiture Controversy.	1121 *Ibn Tumart*, claiming to be Mahdi, preaches puritanical Islam and founds *Almohades* dynasty in N. Africa.	
		1126 Song (Sung) capital *Kaifeng* in China sacked by Jin horsemen; retreat south to *Xingsai*.	
1135	Civil war in England: *Stephen* against *Matilda*.		*Abelard* (1079–1142) teaching in Paris.
1137	*Catalonia* linked by marriage with *Aragon*.		1136 Geoffrey of Monmouth: *History of the Kings of Britain* (Arthur).
1138	Henry of Bavaria of *Guelph* family disputes crown of Germany with *Conrad III* of Ghibellines; beginning of long medieval struggle: *Guelph* (for pope) against *Ghibelline* (for emperor).	*Anasazi* Indians in Colorado build *Mesa Verde*.	1140 Bernard obtains *condemnation of Abelard*.
			Gothic architecture; St Denis west front.
1147	The *Almohades* rule in southern Spain.	1147–8 *Second Crusade*; achieves little.	c.1150 Toledo school of translators transmits Arab learning to West.
1152	*Henry of Anjou* marries *Eleanor of Aquitaine*.	c.1150 *Khmer* temple *Angkor Wat* built in Cambodia.	c.1150 *Paris* University.
1154	*Henry II* king of England (1154–89).		
1155	*Frederick I* ('*Barbarossa*') crowned Holy Roman Emperor by Pope Hadrian IV (1155–90); seeks to extend power in Italy.	*Toltec* capital *Tula* overrun by *Chichimecs* from N. Mexico. Toltec power *declines*.	1158 University of *Bologna* granted charter by Frederick I.
1165	*William* ('*the Lion*') king of Scotland (1165–1214).		
1166	*Assize of Clarendon* establishes *jury system* in England.	1168 *Aztecs* moving into Mexico; destroy Toltec Empire.	1167 *Catharist* Manichaean heresy spreading.
		1169 *Saladin* (1137–93) founds dynasty of *Ayyubids* and conquers Egypt. Drives Christians from Acre and Jerusalem.	1167 *Oxford* University.
1170	Murder of *Thomas à Becket*.		c.1170 *Tristan et Iseult*.
			Troubadour songs in France.
			Early *polyphonic* music.
1173	Henry II's sons rebel.		1171 *Averroës* teaching in Cordoba.
1177	*Peace of Venice*; Frederick Barbarossa accepts that cardinals elect pope.		
1179	Third *Lateran Council* condemns *Catharism*, authorizes crusade against Albigensians.		

CHRONOLOGY OF WORLD EVENTS

	Britain and Europe	Rest of the World	Culture/Technology
1180	*Philip II ('Augustus') king of France (1180–1223); greatly expands kingdom.*	1181 *Khmer Empire at height, under Jayavarman VII.*	*Longbow* in Wales.
1182	*Massacre of Latin merchants* in Constantinople.	1185 Japanese shogunate of *Minamoto Yoritomo* at Kamakura.	
1189	*Richard I king of England (1189–99).*	*Third Crusade; Acre retaken (1191). Saladin grants pilgrims access to Holy Places.*	1193 *Zen Buddhism* in Japan.
1198	*Innocent III pope (1198–1216): papacy has maximum authority during these years.*		1194 Chartres Cathedral rebuilding begun in Gothic style. Chartres windows.
1199	*John king of England (1199–1216).*	1200 *Incas developing civilization based on Cuzco.* Post-Classical civilization in Peru (*Chimu*, 1200–1465); large urban centres; elaborate irrigation.	
1204	*Fourth Crusade; Constantinople sacked.* *Philip Augustus victory in Normandy.*	1206 *Genghis Khan (1162–1227) proclaims Mongol empire.* 1206 Muslim *Mameluke sultanate* in Delhi established.	1202 *Arabic mathematics* in Pisa (Fibonacci). *Louvre* built (fortress of Philip Augustus).
1212	*Children's Crusade*; thousands enslaved. Battle of *Las Navas de Tolosa* in Spain: *Moors ejected from Castile.*		1209 *Cambridge* University. Franciscan (1209) and Dominican (1216) Orders.
1214	*Battle of Bouvines:* Philip Augustus against John and German emperor Otto IV. Philip gains *Normandy, Maine, Anjou,* and *Poitou* for France.		
1215	John accepts *Magna Carta* from barons.	Beijing (Peking) sacked and Jin empire destroyed by *Mongols.*	Islam spreading into SE Asia and Africa.
1216	*Henry III* of England aged 9 (1216–72); *William Marshall* regent until 1227.		
1220	*Frederick II emperor (1220–50); inherits S. Italy and Sicily, which he makes power-base.*		
1223	*Louis VIII of France (1223–6) conquers Languedoc in Albigensian Crusade against Cathars (1224–6).* *Mongols invade Russia.*		1224 Frederick II founds *Naples University:* Jews, Christians, and Arabs. Frederick's court at *Palermo* and *Lucca* centre of Byzantine and Arab culture.
1226	*Louis IX of France (1226–70).*	1229 *Frederick II negotiates access for pilgrims to Jerusalem, Bethlehem,* and *Nazareth.*	
1236	*Cordoba* falls to Castile.	c.1235 Sundjata conquers Ghana and establishes *Mali empire* in West Africa.	c.1236 *Roman de la Rose.*

	Britain and Europe	Rest of the World	Culture/Technology
1241	Formation of Hanseatic League.	Mongol *'Golden Horde'* khanate.	
1250	*Italian cities* gain power on collapse of Frederick II's empire.	*Ayyubid dynasty* falls to Mongols and Christians are driven from Jerusalem.	1248 *Alhambra* begun. 1248 *Cologne* Cathedral begun.
		Mongol Mamelukes establish dynasty in Egypt.	
1258	*Provisions of Oxford* limit royal power in England.	Mongols take *Baghdad.*	Thomas Aquinas (1225–74): *Summa theologiae.*
	Catalans under *James I of Aragon (1213–76) expel Moors* from Balearics.		*Tin plate* for armour (Bohemia) and *screw-jack* (1271 in sketchbook of Villard de Honnecourt).
1261	Byzantines regain Constantinople.	1259 *Kublai Khan* elected Great Khan. Establishes capital in Beijing and founds *Yuan dynasty.*	Artists move from Apulia to Italian cities: *Nicola Pisano, Pulpit c.*1260; renaissance of classical style.
1264	*Battle of Lewes;* Simon de Montfort effective ruler of England until *Evesham* in 1265.		
1270	*Louis IX (St Louis)* dies on Crusade outside *Tunis. Philip III* succeeds (1270–85).	Marco Polo travelling 1271–95.	*Gothic architecture* throughout Europe.
1272	*Edward I* of England (1272–1307) begins conquest of *Wales.*		
1282	*Sicilian Vespers;* revolt against Angevins (ruled since 1266). Sicily goes to *Aragon.*		*Duccio* (d. 1319) in Siena.
	Surrender of Harlech castle: Welsh revolt collapses.		
1284	*Statute of Rhuddlan;* English rule of Wales confirmed.		*Giotto* (c.1267–1337) establishes modern painting in Florence.
1285	*Philip IV* of France (1285–1314).	*Inca empire* expanding in Peru.	
1290	*Jews* expelled from England.		
	Scottish throne vacant.		
1297	*William Wallace* defeats English army at *Stirling Bridge.*	New *empire of Maya* flourishing in Yucatán (Chichén Itzá): stone architecture; pottery; gold artefacts; elaborate temples.	Roger Bacon: *Opus maius* (1266); imprisoned for heresy.
1305	Edward I *executes* Wallace.		1303 *Spectacles* invented.
	Clement V pope (1305–14); papacy moves to *Avignon* (1309).		
1306	*Jews* expelled from France.		*Duns Scotus* (c.1260–1308) and *nominalists* oppose Aquinas' theology.

	Britain and Europe	Rest of the World	Culture/Technology
1307	*Knights Templar* suppressed in France. *Edward II* king of England (1307–27). *Italian cities* flourish as German empire abandons control.	*Empire of Benin* emerging in southern Nigeria.	Dante: *Divina Commedia* begun c.1307.
1314	*Robert Bruce* defeats English at Bannockburn.		
1327	Edward II imprisoned and *murdered*. *Edward III* of England (1327–77); his mother *Isabella* and her lover *Mortimer* rule till 1330.	c.1325 *Tenochtitlán* founded by Aztecs; cult of *Quetzalcóatl* from Toltecs. *Ibn Batuta* (c.1304–68) travelling.	*Spinning-wheel* from India in Europe.
1328	Scottish *independence* recognized. Capet line of kings ends. *Philip VI* first *Valois* king of France (1328–50).	c.1330 Disease (plague) and famine weaken *Yuan dynasty*.	1329 *Meister Eckhart* (d. 1327) condemned posthumously.
1338	*Edward III* claims French throne and Hundred Years War begins.	1336 *Vijayanagar* city founded S. India; seat of Hindu empire till 1565. 1336 In Japan *Ashikaga* shogunate founded.	1335 *Mechanical clocks* at Milan and Wells.
1340	*Battle of Sluys*; English gain control of Channel.		*Paper-mill* at Fabriano. *Bruges* centre of *wool trade*; *Flemish art* emerging.
1346	*Battle of Crécy*: English victory; cannon used; Calais occupied.		Petrarch: *Canzoniere*; crowned *poet laureate* 1341. c.1344 *Order of Garter* in England.
1348	*Black Death* arrives in England; one-third of population dies.	1348 *Black Death* reaches Europe from China.	Boccaccio: *Decameron*.
1351	English *Statute of Labourers* seeks to *uphold* feudalism.	*Aztec Empire thriving*: gold; copper; obsidian; calendar and hieroglyphic writing; mathematics based on 20 with zero; *Tenochtitlán* city of *300,000*; pyramids with elaborate rituals with *human sacrifice*.	
1353	*Ottoman* Turks enter Europe.		
1356	Edward *Black Prince* wins *Poitiers*; French king *John II* captured.		
1358	Revolt of *Étienne Marcel* in Paris.		
1360	*Treaty of Brétigny*; England keeps western France, in peace with French.		1363 *Guy de Chauliac* advances medicine from Black Death studies. 1364 Machaut: *Mass of Nôtre Dame*. William Langland: *Piers Plowman*.

	Britain and Europe	Rest of the World	Culture/Technology
1369	Hundred Years War resumed; French under Du Guesclin.	Mongol Tamerlane (Timur the Lame) conquers Turkestan, Delhi, Persia, Golden Horde, Syria, and Egypt (1363–1405).	Jean Froissart: Chronicles. Siena artists flourish.
		Ming dynasty in China founded by Zhu Yuanzhang (Chu Yuan-chang) (1368–1644); great period for pottery and bronze.	
1371	Robert II king of Scotland (1371–90); first Stuart.		
1377	Richard II king of England (1377–99). Papacy returns to Rome from Avignon.		
1378	Great Schism in Church; two popes.		Flamboyant architecture in Europe: Beauvais Cathedral
1380	Charles VI king of France (1380–1422).		
1381	Peasants' Revolt in England; Wat Tyler defeated; poll tax withdrawn.		1382 Lollards (John Wyclif) condemned.
1386	Poland and Lithuania unite.	1392 Yi dynasty in Korea (1392–1910) under Ming influence.	1387 Chaucer: Canterbury Tales.
1396	Truce in Hundred Years War. Battle of Nicopolis; Crusaders in Hungary defeated by Turks.		Ghiberti in Florence (1378–1455): Baptistery doors. c.1389 Sluter in Dijon: Well of Moses.
1397	Union of Kalmar: crowns of Denmark, Norway, and Sweden unite (1397–1523).		
1399	Richard II deposed and murdered; Bolingbroke Henry IV 1399–1413.		
1402	Owen Glendower defeats English at Pilleth.	Timur defeats Ottomans at Ankara. Foundation of Malacca by Srivijaya; becomes entrepôt.	
1403	Prince Henry ('Hal') defeats Percy ('Hotspur') and Glendower rebellions (1408).	Zheng He's (Cheng Ho's) voyages; reaches Persian Gulf and E. Africa from China (1405–33).	Metal screws in Europe.
1413	Henry V king of England (1413–22).	Portuguese voyages begin under Henry the Navigator (1394–1460).	St Andrews University. Duc de Berry: Très riches Heures.
1415	Henry wins Agincourt; occupies Normandy. Council of Constance condemns John Huss to stake; ends Schism.		Hussites seek revenge for martyr Huss. Painting in oil begins.

	Britain and Europe	Rest of the World	Culture/Technology
1422	Henry VI king of England (infant); Council of Regency claims France.		Masaccio (d. 1428) frescoes.
	Charles VII king of France (1422–61).		1420 Dome of Florence Cathedral begun by Brunelleschi.
1429	Joan of Arc relieves Orleans.		
1431	Joan burnt at stake in Rouen.	Thais of Siam take Angkor. Phnom Penh new Khmer city.	
1434	Cosimo de' Medici rules in Florence; patron of learning.	1438 Inca ascendancy in Peru; high level of astronomical and surgical knowledge; cotton and potato grown.	Donatello: David.
	Burgundy emerging under strong dukes.		Van Eyck: Arnolfini portrait.
		1441 Maya city of Mayapán conquered by Uxmal.	Thomas à Kempis (d. 1471): Imitation of Christ.
1450	Jack Cade's peasant rebellion suppressed.		Nicholas of Cusa (d. 1464): astronomy; theology; revives Neo-Platonism.
1452	Frederick III (1452–93) first Hapsburg Holy Roman Emperor.		1452–66 Piero della Francesca: Legend of Cross (Arezzo).
1453	Constantinople falls to Ottoman Turks.	Hindu Majapahit empire in Java declining; Islam advancing.	1456 Alberti: façade of Santa Maria Novella; Florence centre of artistic activity.
	Hundred Years War ends. Henry VI 'insane'; Richard duke of York protector; English 'Wars of Roses' begin.		
1460	Richard of York killed at Wakefield.		c.1456 Gutenberg Bible. c.40,000 editions printed 1450–1500.
1461	Edward of York seizes throne; Edward IV (1461–83).		1456–60 Uccello: Battle of S. Romano.
	Louis XI king of France (1461–83).		Lace in France and Flanders.
1462	Ivan III of Russia rejects control of Horde.		
1469	Ferdinand and Isabella marry; unite Aragon and Castile (1479).		
1471	Lancastrians defeated at Tewkesbury; Henry VI killed; Edward IV accepted (1471–83).		
1477	Death of Charles the Bold of Burgundy.	1476 Incas conquer Chimú.	1474 Caxton printing at Westminster.
1480	Ivan III overthrows Mongol Golden Horde.		1478 Topkapi Palace in Constantinople.
			c.1478 Botticelli: Primavera.
1483	Edward V; Richard III king of England (1483–5).		
1485	Henry Tudor wins Bosworth; Henry VII (1485–1509).		Leonardo (1452–1519): anatomy; mechanics; painting, etc.

	Britain and Europe	Rest of the World	Culture/Technology
1492	*Reconquest* of Spain complete.	*Columbus* reaches *West Indies.*	
		1493 *Askia Muhammad* emperor of *Songhay* on Niger, West Africa.	
1494	*Treaty of Tordesillas* divides New World between Spain and Portugal.	1497 *John Cabot* reaches mainland North America from Bristol.	Nanak (1469–1539) founds the Sikh religion.
	Charles VIII invades Italy.		
1498	*Louis XII* king of France (1498–1515).	1499 Vasco da Gama rounds Cape and reaches *Calicut*; beginning of *Portuguese empire.*	
1502	Death of *Arthur Tudor* in England.	*Safavid dynasty* in Iran: Ismael I (1501–24).	1501 Michelangelo carves *David.*
	Margaret Tudor marries James IV of Scotland.		
1503	*Julius II* pope (1503–13).		c.1504–5 Leonardo paints *Mona Lisa.*
			1505–7 *Dürer* in Italy.
			1506 Bramante designs *St Peter's,* Rome.
1509	*Henry VIII* king of England (1509–47).		1508–12 Michelangelo: *Sistine ceiling.*
			1509 *Watch* invented in Nuremberg.
1513	Scots defeated at *Flodden*; death of James IV.	1511 Portuguese conquer *Malacca.*	
1515	*Francis I* king of France (1515–47).		1516 Grünewald: *Isenheim Altar.*
			1516 King's College Chapel, Cambridge completed.
1517	Start of *Protestant Reformation,* Germany.	Ottoman Turks conquer *Egypt.*	*Erasmus'* last visit to England.
1519	*Charles V* (Hapsburg) elected Holy Roman Emperor.	*Cortés* conquers *Aztecs.*	
		Magellan crosses Pacific.	
1521	*Diet of Worms* condemns Luther's teaching.	*Suleiman the Magnificent* sultan of Turkey (1520–66).	1522 Ignatius Loyola: *Spiritual Exercises* (pr. 1548).
1524	*Peasants' War* in Germany.		
1525	France loses control of N. Italy; *Battle of Pavia.*	c.1525 *Babur* from Kabul invades India and founds *Mughal dynasty.*	
	Reformation moves to Switzerland.		
1526	Battle of *Mohács.* Ottoman Turks occupy Hungary.		
1527	Charles V's troops *sack Rome.*		*Paracelsus* in Basle (new concept of disease).
1529	Ottomans *besiege Vienna.*	Franciscan mission to Mexico.	Early Italian *madrigal.*
	Fall of English chancellor *Wolsey.*	European *spice trade* with Asia; *sugar/slaves* with America.	

Britain and Europe	Rest of the World	Culture/Technology
1533	Henry VIII marries *Anne Boleyn*.	*1531–3 Pizarro* conquers *Inca empire*. Atahualpa killed.	
1534	English *Act of Supremacy*; break with papacy.	*Iran* conquered by Ottoman Turks.	*Luther's Bible.*
	Anabaptists revolt in Munster, Germany.		Rabelais: *Gargantua.*
			Jesuits founded.
1535	*John Calvin* in Geneva.		
1536	*Dissolution* of English and Welsh *monasteries*.		Calvin: *Institutes.*
1540	Henry VIII tries to impose political and religious settlement on *Ireland*.	*1542 Francis Xavier* in India, Sri Lanka, Japan (1549).	1543 Copernicus: *De revolutionibus.*
			1543 Vesalius: *De humani corporis fabrica.*
1545	*Council of Trent* begins (1545–63).		Calvin: *Letter on Usury.*
1547	*Edward VI* king of England (1547–53).	Portuguese settling coast of Brazil.	
	Ivan the Terrible crowned tsar (1547–84).		
1553	*Mary Tudor* queen of England (1553–8); persecutes Protestants.		1550 Vasari: *Lives of the Most Excellent Painters etc.*
1555	*Peace of Augsburg*: pacification of Germany; Calvinists persecuted.		
1556	Charles V retires; *Philip II* king of Spain (1556–98).	*Akbar* emperor of India (1556–1605). Expands empire and unites its peoples.	
1558	France recaptures *Calais* from English.	1557 Portuguese found *Macao*.	
	Elizabeth I queen (1558–1603).		
1559	*John Knox* active in Scotland.		*Tobacco* enters Europe.
1562	Start of French *Wars of Religion*.		
1563	*Religious settlement* in England; *39 Articles*.		
1567	*Dutch Revolt* begins.		1568–71 Palladio builds *Villa Rotonda* in Vicenza.
	Mary Queen of Scots flees to England.		
1570	Pope excommunicates Elizabeth I.		1569 Mercator invents *map projection*.
			1569 Death of *Brueghel*.
1571	*Battle of Lepanto*; Turkish domination of Mediterranean ends.		
1572	*Massacre of St Bartholomew* (French Huguenots).	1573 *Oda Nobunaga*, Japanese warrior, imposes political order on *Japan*.	Camoens: *Os Lusíadas*.

Britain and Europe	Rest of the World	Culture/Technology
1577 Drake's *voyage round the world* begins.	1576 Last Hindu kingdoms fall to Akbar, who welcomes Jesuits. Spanish expanding *New Spain* and *New Granada* in America.	1576 Titian: *Pietà*. *El Greco* to Toledo. Tycho Brahe: *De nova stella*.
1579 *Union of Utrecht* unites Protestant Dutch. *Irish* rebels massacred; English *plant settlers*.		1580–95 Montaigne: *Essais*.
1580 Spain occupies *Portugal*.		
1581 English *Levant Co.* founded.	1582 Warrior Hideyoshi *unites Japan* and campaigns in Korea (1592, 1598).	
1585 *War of Three Henries* in France. England and Spain *at war*.	1584 *Walter Raleigh* founds colony of *Virginia*.	
1587 Mary Queen of Scots *executed*.		
1588 *Spanish Armada* defeated. *Duke of Guise* murdered in France.		
1589 *Henry of Navarre* claims throne and besieges Paris.	1591 *Morocco* conquers Islamic kingdom of *Songhay* on River Niger.	Early *ballet* in France. c.1590 Marlowe: *Faustus*. 1590, 1596 Spenser: *Faerie Queene*.
1593 *Henry IV* (1589–1610) accepts *Catholicism* in France.		1592 *Monteverdi* to Mantua. Early *microscope; thermometer; water-closet*.
1594 Bad harvests in England (1594–7).		Death of *Palestrina* and *Lasso*.
1598 *Edict of Nantes* ends Wars of Religion in France. *Boris Godunov* Russian tsar (1598–1605).		1596 Shakespeare: *Romeo and Juliet*.
1599 *Irish revolt* suppressed by Essex and Montjoy (1601).		*Globe Theatre* built.
1600 Rebellion of *Essex* against Elizabeth.	English *East India Co.* formed.	*Giordano Bruno* burnt for theory of universe. 1600–8 *Rubens* in Italy.
1603 *James VI* of Scotland and *James I* of England (1603–25).	*Tokugawa* shogunate established in Japan.	1602 Shakespeare: *Hamlet*.
1604 Anglo-Spanish *Peace Treaty*.		
1605 *Gunpowder Plot* in English Parliament.	1607 *Virginia* settled by British.	1605–6 Ben Jonson: *Volpone*. 1605, 1615 Cervantes: *Don Quixote*. 1607 First *opera* in Mantua: Monteverdi, *Orfeo*.
1609 *Dutch Republic* recognized.	1608 *Quebec* settled by Champlain for France.	Shakespeare: *Sonnets*. Galileo: *telescope*. Kepler: *Laws of Planetary Motion*.

	Britain and Europe	Rest of the World	Culture/Technology
1610	Ulster planted with English and Scottish settlers. Louis XIII king of France (1610–43).		Caravaggio dies (1571–1610). 1611 Authorized Version of the Bible. Shakespeare: Tempest.
1613	Russian 'Time of Troubles' ends; Romanov dynasty established. Elizabeth Stuart marries Elector Palatine.		1614 John Napier work on logarithms.
1618	Thirty Years War begins in Europe.	1616 Japan ejects Christian missionaries. Tobacco plantations in Virginia expanding.	1616–21 Inigo Jones designs Banqueting House. 1618 French salon established under Marquise de Rambouillet.
1620	Emperor Ferdinand II wins battle of White Mountain in Bohemia.	Pilgrim Fathers arrive at Cape Cod on Mayflower.	F. Bacon: Novum organum.
1621	Philip IV king of Spain (1621–65).		
1624	Richelieu in power in France. Britain and Spain renew war (1624–30).		
1625	Charles I king of England (1625–49).		
1626	Count Wallenstein leads Imperial armies in war.	Dutch purchase Manhattan (New Amsterdam).	
1627	Britain and France at war 1627–9. Richelieu defeats Huguenots.		
1629	Charles I governs without Parliament.	1630–42 Large-scale British emigration to Massachusetts.	1628 W. Harvey: De motu cordis.
1630	Gustavus Adolphus of Sweden joins Thirty Years War and is killed at Lützen (1632).		1632 Van Dyck to England.
1633	William Laud elected Archbishop of Canterbury; opposes Puritans in Britain.		Galileo before Inquisition; recants.
1634	Wallenstein assassinated.		
1635	France joins Thirty Years War.		
1637	Charles I faces crisis in Scotland over new liturgy.		Corneille: Le Cid. Descartes: Co-ordinate geometry and Discours de la méthode. Waterproof umbrellas used at court of Louis XIII of France.
1639	France occupies Alsace.	Japan closed to all Europeans.	
1640	Long Parliament begins. Portugal regains independence from Spain.		
1642	First English Civil War begins. Death of Richelieu.	1642–3 Tasman explores Antipodes.	Rembrandt: Night Watch. Pascal's calculating machine.

	Britain and Europe	Rest of the World	Culture/Technology
1643	*Louis XIV* king of France (1643–1715). *Mazarin* in power.		Torricelli invents *barometer*.
1644	Charles I defeated at *Marston Moor*.	*Qing (Ch'ing)* dynasty established in China.	
1645	*New Model Army* formed. Charles defeated at Naseby. War ends (1646).		
1647	*Leveller* influence in Army.		
1648	*Second Civil War.* King accepts defeat; tried and *executed* (1649). *Peace of Westphalia* ends Thirty Years War.	Atlantic *slave trade* expanding.	*c.*1648 *Taj Mahal* completed.
1649	*Cromwell* massacre at *Drogheda*, Ireland.		
1652	*First Anglo-Dutch War.* Cromwell conquers *Scots*.	1652 Dutch found *Cape Colony*.	1650 *Air-pump* (Germany). *c.*1650 Poussin: *Shepherds of Arcadia*. 1651 T. Hobbes: *Leviathan*.
1653	Oliver Cromwell *'Protector'* (1653–8).		
1654	*Queen Christina* of Sweden abdicates.		1656 Bernini completes piazza of St Peter's, Rome.
1658	*Peace of Roskilde* and *Treaty of Pyrenees* (1659) end period of war in Europe.	*Aurangzeb* Mughal emperor (1658–1707); expansion followed by decline after his death.	1656 Huygens invents *pendulum clock*. *c.*1656 Velázquez: *Las Meninas*.
1660	*Charles II* restored as king (1660–85).		*Vermeer* at work.
1661	Louis XIV begins *personal reign*. *Colbert* minister (1665–83).		1662 *Royal Society* in London. 1662 R. Boyle: *Boyle's law*.
1664	*Second Anglo-Dutch War* (1664–7).	Colbert founds *French East India Co.*	Molière: *Tartuffe*. Frans Hals: *The Regents*.
1665	*Great Plague* of London.		
1666	*Great Fire* of London.	1667 Hindu *Maratha kingdom* founded by *Sivaji* in western India; challenges Mughals.	*Academy of Science* in France. 1666–7 Newton invents *differential calculus*. *Spirit-level* invented. 1667 Milton: *Paradise Lost*.
1670	French troops occupy *Lorraine*.	Rise of *Ashanti* in W. Africa.	1668 *Versailles*; palace rebuilt by Le Vau, Le Brun, and Le Nôtre. 1670–1720 Wren rebuilds London churches and St Paul's.
1672	*Third Anglo-Dutch War* (1672–4).	1675 In India *Sikhism* becomes *military theocracy* to resist Mughal power.	1671 Mme de Sévigné: *correspondence*.

	Britain and Europe	Rest of the World	Culture/Technology
1678	*Franche-Comté* to France.		1676 *Van Leeuwenhoek* finds microbes with the aid of *microscope.*
			1677 Racine: *Phèdre.*
			1678 Mme de Lafayette: *Princesse de Clèves.*
			1678–84 Bunyan: *Pilgrim's Progress.*
1682	*Peter the Great* tsar of Russia (1682–1725).	1681 *Pennsylvania* founded. *Carolinas* flourish.	1681 *Pressure-cooker* invented.
1683	Turks under Kara Mustafa *besiege Vienna.*		
1685	*Edict of Nantes* revoked.		
1687	Hapsburgs recover *Hungary.*		Newton: *Principia Mathematica.*
1688	English '*Glorious Revolution*'; William (1689–1702) and Mary (1689–94) reign.		
1689	War with France (War of *League of Augsburg*).	*Treaty of Nerchinsk* between China and Russia, fixing frontiers.	Purcell: *Dido and Aeneas.*
1690	Battle of the *Boyne*; Irish and French defeated. French fleet sunk.	1692 *Witch trials* in *Salem.*	Locke: *Essay concerning Human Understanding.*
1697	*Treaty of Ryswick* ends War of League of Augsburg.		1693 *François Couperin* to Versailles as court organist.
			1698 T. Savery: *steam engine.*
1700	*Charles II* (Hapsburg) of Spain dies childless.		Congreve: *Way of the World.* *Stradivarius*: violins.
	Philip V (Bourbon) king of Spain (1700–46).		
	Great Northern War (1700–21). Russians defeated by Sweden.		
1701	*War of Spanish Succession* (1701–13): Britain, Netherlands, Austria, German princes, Savoy, and Portugal against France, Bavaria, and Castile.		
1702	*Queen Anne* (1702–14).		
1704	*Marlborough* wins battle of *Blenheim.*		Newton: *Optics* (explains colour).
	British take *Gibraltar.*		1705 *Halley* predicts return of his comet.
1706	Battle of *Ramillies*: Marlborough routs French.		
1707	*Act of Union* between England and Scotland.		
1708	Battle of *Oudenarde*: Marlborough's third victory.		J. S. *Bach* at Weimar.

	Britain and Europe	Rest of the World	Culture/Technology
1709	Battle of *Poltava* in Northern War ends Swedish hegemony in the Baltic. Last of *Marlborough's* victories at *Malplaquet*.	1710 British capture French Acadia (*Nova Scotia*).	First *pianoforte* in Italy. 1710 *Jansenism* persecuted in France.
1711	Queen Anne persuaded to *dismiss Marlborough*. Ministry led by *Bolingbroke* who seeks to end war.		*Newcomen* piston-operated *steam engine* in England. 1712 *Handel* to London.
1713	*Treaty of Utrecht*. Philip V confirmed in Spain. Southern Netherlands, Milan, Naples, and Sardinia to Austria. Britain gains *Assiento* to supply *slaves* to Spanish colonies. *Frederick William I* king of Prussia (1713–40).	*Newfoundland, St Kitts*, and *Hudson Bay* to Britain.	Bull *Unigenitus* against Jansenists.
1714	*George*, Elector of Hanover, great-grandson of James I, King George I of England and Scotland (1714–27).		*Fahrenheit* devises mercury thermometer.
1715	*Louis XV* king of France (1715–74); Philippe d'Orléans regent. *Jacobite rebellion* suppressed in Scotland and England.	1717 *Shenandoah Valley* settled. Indians evicted.	1716 F. Couperin: *Treatise on harpsichord playing*. 1717 Watteau: *Embarquement pour Cythère*.
1720	*South Sea Bubble*; major financial collapse in London.	Chinese invade *Tibet*.	1719 Defoe: *Robinson Crusoe*.
1721	R. *Walpole* first effective *Prime Minister*, as king often in Hanover (1721–42). Policy of peace and commercial expansion.	French and English *East India Cos.* rivals in India.	
1726	*Cardinal Fleury* chief minister in France (1726–43).		1724 *Bourse* opens in Paris. 1726 Swift: *Gulliver's Travels*. 1726 *Voltaire* liberated from Bastille; goes to England.
1727	Peaceful succession of *George II* (1727–60).	1728 Danish explorer *Bering* discovers Straits. 1729 *North and South Carolina* become Crown Colonies.	1728 Gay and Pepusch: *Beggar's Opera*. 1728 Pope: *Dunciad*. 1729 Bach: *St Matthew Passion*.
1733	*War of Polish Succession* (1733–8). Russia imposes *Augustus III*, elector of Saxony, on Poland.	1732 *James Oglethorpe* founds *Georgia* for 'poor debtors'.	1732 *Trevi Fountain* in Rome. 1733 J. Kay invents *flying shuttle* for weaving (England). 1734 *Fire extinguisher* invented by German physician M. Fuches. 1734 Voltaire: *Lettres philosophiques*. 1735 C. Linnaeus: *Systema naturae*. 1735 Rameau: *Les Indes galantes*.

Britain and Europe	Rest of the World	Culture/Technology
1739 *War of Jenkins' Ear* between England and Spain.	1738 Sea Captain *R. Jenkins* advocates war in Caribbean against Spain.	1738 *Wesley brothers* 'converted'.
	1738 In W. Africa Yoruba kingdom of *Oyo* conquers Dahomey.	
1740 *Frederick II* king of Prussia (1740–86); claims *Silesia*, causing *War of Austrian Succession*: Austria, Britain, and Hanover against France, Spain, and Prussia.	1741 *Bering* discovers *Alaska*.	Richardson: *Pamela*.
	1741 *Dupleix* commandant-general for French in India.	1742 *Celsius* devises centigrade scale.
		1742 Handel: *Messiah*.
Empress Maria Theresa, queen of Hungary and Bohemia, wife of Francis I (1740–80).		
1743 King George II defeats French at *Dettingen*.		
1745 *Jacobite Rebellion* under 'Bonnie Prince Charlie'; ruthlessly suppressed (1746).	1744–8 *King George's War* in America along St Lawrence.	*Bow Street Runners* in London.
French king's mistress the Marquise *de Pompadour* influences tastes (*rococo*) and policies.		
Treaty of Dresden confirms Frederick II in Silesia.		
1746 *Battle of Culloden* in Scotland: defeat of Jacobites.	*Madras* taken but returned to Britain after war.	
1747 British win naval battle of *Belle-Isle*.	*Afghanistan* united.	*Sans-souci* palace in Prussia.
1748 *Treaty of Aix-la-Chapelle* ends war in Europe.		*Pompeii* excavated.
		Montesquieu: *Esprit des lois*.
		1749 Fielding: *Tom Jones*.
1750 Pombal, reformist minister, in power in Portugal (1750–82).	*East India Companies* (English, French, Dutch) trading extensively in Asia: tea becomes fashionable drink in Europe.	Death of J. S. Bach.
Spirit of *Enlightenment* (Voltaire, Diderot, Montesquieu) influencing all of Europe.		Symphonic form emerging in music.
1751 Death of *Frederick*, prince of Wales.		*Diderot* publishes vol. 1 of *Encyclopédie*.
1752 Britain adopts Gregorian calendar, leading to rioting.		*Franklin* devises *lightning conductor* (used kite to show electrical nature of lightning).
1753 Louis XV exiles *Parlement de Paris*.		*Place de la Concorde* in Paris.
Kaunitz Austrian chancellor (1753–92).		Jewish naturalization in Britain.
1754 Duke of Newcastle leads British ministry.		Buffon: *Histoire naturelle* (36 vols., 1749–88; revolutionizes thinking on animal kingdom).

	Britain and Europe	Rest of the World	Culture/Technology
1755	*Lisbon* earthquake.	*Braddock* expedition in North America against French and Indians; fails to take Fort Duquesne.	S. Johnson: *Dictionary*. *Neoclassical* art fashionable.
1756	Outbreak of *Seven Years War*: Britain, Hanover, Prussia, and Denmark against France, Austria, Russia, and Sweden. R. *Pitt* (the Elder, later *Earl of Chatham*) joins ministry.	*Rivalry* developing in India: French against British.	
1757	*Frederick II* wins at *Rossbach* and *Leuthen*.	*Robert Clive* commands East India Co. army. *Battle of Plassey*: Clive defeats Nawab of *Bengal* and controls the State.	*Sextant* designed by J. Campbell (England). French *physiocrats* exalt Nature and the Land. Stimulate *agricultural reform* in Britain and France.
1758	*Choiseul* first minister in France. Frederick II defeats Russia at *Zorndorf*.		Helvétius: *De l'Esprit* (book burnt for atheism).
1759	'*Annus mirabilis*': French defeated by Ferdinand of Brunswick at *Minden* and by Admiral Hawke at *Quiberon Bay*.	*Wolfe* captures city of *Quebec* and British then take *Montreal* and all colony of *Quebec*.	1759–67 L. Sterne: *Tristram Shandy*.
1760	*George III*, grandson of George II, king (1760–1820).		
1761	Chatham resigns; *Bute* British PM (1762–3).	*Eyre Coote* takes *Pondicherry* from French in India; later returned.	*Haydn* to court of Esterházy in Hungary.
1762	Accession of *Catherine the Great* in Russia, deposing her husband Peter III; *empress 1762–96*; influenced by Diderot and Voltaire and belief in 'Enlightened Despotism'. Russian landowners relieved of military service.	African *slave trade* begins to attract criticism.	J.-J. Rousseau: *Émile* and *Contrat social*. Gluck: *Orphée et Euridice*.
1763	*Peace of Paris* ends Seven Years War: France cedes Canada, the Mississippi, and control of India to Britain. J. *Wilkes* arrested for libel while an MP; expelled from Commons 1764.	American colonists move west into *Ohio* basin. Fort Duquesne becomes *Pittsburgh*.	*Compulsory education* in Prussia.
1764	Jesuits expelled from France.		Voltaire: *Dictionnaire philosophique*.
1765	Rockingham British PM. *American Stamp Act* to finance defence of colonies. *Joseph II* Hapsburg emperor, supported by mother *Maria Theresa*.	Clive recalled to *govern Bengal*; accused of corruption (1772).	1765–70 J.-J. Rousseau: *Confessions*.

	Britain and Europe	*Rest of the World*	*Culture/Technology*
1766	*Stamp Act repealed.* Chatham briefly PM. *Lorraine* integrated into France.	1766–9 *Bougainville's* voyage round world, exploring many Pacific islands.	*H. Cavendish* isolates hydrogen.
1767	Catherine publishes *Instructions*, westernizing Russian law.	Increasing *opposition* in British American colonies to control from London through royal governors. W. African kingdom of *Benin* declining in power.	A. Young: *Farmer's Letters.*
1768	*Corsica* purchased by French.	1768–71 *Cook's* voyage in *Endeavour*; charts *New Zealand* and eastern *Australia*.	Robert and William *Adam* architects: Adelphi Terrace. Wright of Derby: *Experiment with an Air-Pump.*
1769	Russia advancing against Turks and Tartars. *Moldavia, Wallachia*, and *Crimea* occupied in war (1769–74). Birth of *Napoleon*.		Improved *steam engine* using condenser patented by J. *Watt*.
1770	*Lord North* ministry in Britain (1770–82). *Crime* increasing in London; Bow Street Runners developing (formed c.1745).	Policies of *North* government cause growing *resentment* in Virginia and Massachusetts. 'Boston Massacre' (5 die).	Edmund Burke: *Thoughts on the Present Discontents.* *Spinning-jenny* patented by J. Hargreaves; multiple spindles powered by water.
1771	Johann Struensee reforms in Denmark. Influence of *physiocrats* results in new attitudes to *agriculture* and *animal husbandry* (Arthur Young).		Gainsborough: *Blue Boy*. French *salons* dictate taste to Europe and America.
1772	First *partition of Poland* between Russia, Prussia, and Austria.		*Horace Walpole* (d. 1797) rebuilds *Strawberry Hill* in Neo-Gothic.
1773	*Russian uprising* to support bogus emperor, the Cossack *Pugachev*. Suppressed by Catherine with great cruelty.	'Boston tea-party' against taxation without representation.	*Iron bridge* at Coalbrookdale. Arthur Young: *Observations on the Present State of the Waste Lands of Great Britain.*
1774	Financial and administrative *chaos* at the end of Louis XV's reign. *Louis XVI* king (1774–93). British Parliament passes *Coercive Acts* against Boston and Massachusetts. Treaty of *Kutchuk Kainarji* recognizes Russian control of *Black Sea*.	*Warren Hastings* governor-general in India; consolidates Clive's conquests.	Goethe: *Sorrows of Young Werther.* J. Priestley discovers *oxygen* ('dephlogisticated air').
1776	*Turgot* French minister; tries to reform but condemned; *Necker* succeeds (1777–81). French support American colonies (*La Fayette*).	1775 Second *Continental Congress*, Philadelphia. American revolution; battle of *Bunker Hill*. *Declaration of Independence* signed 4 July by 11 rebel colonies.	1775 Beaumarchais: *Barber of Seville.* *Jenner* discovers principle of vaccination. 1776–88 Gibbon: *Decline and Fall of the Roman Empire.* 1776 Adam Smith: *Wealth of Nations.*

	Britain and Europe	US and Rest of the World	Culture/Technology
1777	Rapid growth of British *textile industry*.	British take *New York* and *Philadelphia* but are defeated at *Saratoga*. *Valley Forge* Washington's winter camp: 11,000 die.	Sheridan: *School for Scandal*. *Lavoisier* theory of respiration. *Sturm und Drang* literary movement.
1778	*Bath* dominates polite British society. France joins *America* in war. *Evangelical churches* reviving in Britain (Methodism), Germany and Switzerland (Pietism), and America (Great Awakening).	1778 Washington revives hope as *France joins war*.	Mozart: *Paris Symphony*. *La Scala* in Milan.
1779	*Spain* joins American war. *Riots* against machines in England.		
1780	*Gordon Riots* in London (mob incitement against proposed easing of anti-Catholic laws); *Wilkes* helps to suppress.		*Bath Pump Room* rebuilt.
1781	*Joseph II* of Austria seeks to modernize the empire as 'Enlightened Despot'.	British take *Charleston* but *Cornwallis* surrenders at *Yorktown*. French admiral *de Grasse* had supremacy at sea (17 Oct. 1781).	Kant: *Critique of Pure Reason*. Planet *Uranus* discovered by Herschel.
1782	*Political crisis* in Britain. Fall of Lord North. Lord Rockingham briefly PM.		Laclos: *Les Liaisons dangereuses*. Goethe PM Weimar. *Watt* patents double-acting rotary steam engine.
1783	Fox–North Coalition. *Pitt the Younger* PM (1783–1801). *Treaty of Versailles*: Britain accepts independence of 13 Colonies but retains *West Indies* and *North American* Canadian colonies.	Defeated British *loyalists* leave victorious colonies and move to *New Brunswick* and Prince Edward Island.	*Hot-air balloon* (France).
1784	Russia annexes *Crimea* and founds *Sebastopol* (Potemkin). *East India Act* (East India Co.). *Methodists* split from Church of England.		*Cavendish* synthesizes water $H + O$.
1785	Pitt's attempt at *parliamentary reform* defeated. Catherine II grants *Charter* to nobility who exploit serfdom in Russia. *Eden commercial treaty* with France.	Warren Hastings returns to Britain to face 7-year trial for *corruption* in India.	David: *Le Serment des Horaces* (neoclassical). *Power loom* patented by Cartwright (England). 1786 Mozart: *Marriage of Figaro*.
1787			Schiller: *Don Carlos*. Charles formulates *Law*.

	Britain and Europe	*US and Rest of the World*	*Culture/Technology*
1788	George III's first *insanity*. France *bankrupt*; *Necker* restored.	First British convicts to *Botany Bay* under Philip. Settle in Sydney Bay and create *New South Wales*. *US Constitution* agreed. *George Washington* elected first US *President*.	Improved horse-drawn *threshing-machine* patented by Meikle.
1789	French Revolution: *Estates General* summoned (5 May); *Bastille* stormed (14 July); *National Assembly* called; French aristocrats flee to England and Germany. Torture ends in France as liberalization of government attempted. French *corvée* ended.	Washington inaugurated Jan. 1789. President until 1797. *Alexander Hamilton* to US Treasury.	Blake: *Songs of Innocence*. *Lavoisier* establishes *modern chemistry*.
1790	France: *Church* lands nationalized; 83 *French Departments*.	US Supreme Court meets in New York.	Burke: *Reflections on the French Revolution*. *Ambulances* in France.
1791	*Counter-revolutionaries* organize military invasion of France from Germany. Louis XVI and Marie Antoinette *under restraint* in Paris. Attempted escape foiled; arrested at *Varennes*. *Revolutionary army* trained.	Sydney NSW growing with *convict labour*. US *Congress* meets in Philadelphia; selects site of District of Columbia. Slave revolt in *Haiti* under *Toussaint L'Ouverture*.	Mozart dies in poverty. 1791–2 Thomas Paine: *Rights of Man*. *Louvre* becomes a museum.
1792	*Russian empire* extends beyond Black Sea. *French Republic* proclaimed (Sept.); beginning of *Terror*; victory at *Valmy* against Austria.		Mary Wollstonecraft: *Vindication of the Rights of Women*.
1793	*Louis XVI executed* (Jan.). Britain declares *war*; establishes *Board of Agriculture*. *Second partition* of Poland. *Committee of Public Safety* under *Robespierre*. Revolt of *Vendée*; crushed.	George Washington's second term as US President.	
1794	Robespierre *executed* (July).	*Toussaint* defeats Spanish and English force. 'Scottish Martyrs' transported to NSW.	Eli Whitney invents *cotton-gin* in US.
1795	*Third partition of Poland*. Rural depression and high inflation in Britain; *Speenhamland system*: poor relief. France under *Directory*.	'Whisky Rebellion' in US.	*Hydraulic press* in England.
1796	French campaign in Italy; *Bonaparte* victor.	*John Adams* elected US President (1797–1801).	Jenner succeeds with smallpox *vaccine*.

Britain and Europe	US and Rest of the World	Culture/Technology
1797 Talleyrand French foreign minister. *Treaty of Campo Formio* with Austria; *Cisalpine Republic* formed; left bank of Rhine annexed.		Cherubini opera: *Medea*.
British navy mutinies at *Nore*.		
1798 Napoleon to *Egypt*. *Helvetian Republic* formed.	French invade *Egypt*.	*Hansard*: Parliamentary Reports.
Irish rebellion suppressed.		Malthus: *Essay on Population*.
		Lithography invented by Senefelder.
		Wordsworth: *Lyrical Ballads*.
		Coal-gas lighting patented in England.
1799 Napoleon returns to Paris and seizes power as *First Consul*. European coalition against France.	Nelson destroys French fleet at *Aboukir Bay*. French siege of Acre defeated by *Sidney Smith*. Napoleon returns to Paris.	*Gas-fire* patented (France).
Income tax in Britain; *trade unions* suppressed.		
1800 Napoleon defeats Austrians at *Marengo*. France dominates *Italy* except for Sicily and Sardinia.	*Thomas Jefferson* elected US President (1801–9).	*Volta* makes first *battery*.
		Fichte: *The Destiny of Man*.
1801 *Irish Act of Union*.		Chateaubriand: *Atala*.
First *census* in Britain.		Gauss: theory of number.
General Enclosure Act in Britain.		
Concordat between *Napoleon* and *Pope*.		
Alexander I tsar (1801–25).		
1802 *Peace of Amiens*.		*Charlotte Dundas* first steamship.
Napoleon annexes Piedmont.		
1803 Britain declares *war* and forms new coalition.	*Wellesley* defeats Indians in *Maratha War*.	Beethoven: *Eroica Symphony*.
Rebellion in Ireland suppressed.	Jefferson purchases *Louisiana* from Napoleon.	Dalton: *atomic theory*.
1804 *William Pitt* again PM.	*Haiti* independent; Toussaint imprisoned and dies in France.	*Code Civil* in France.
Pope crowns Napoleon *emperor*.	1804–6 *Lewis and Clark Expedition* across US.	*Trevithick*: first steam rail locomotive in S. Wales.
Execution of *Duc d'Enghien* shocks Europe.	*Uthman dan Fodio* leads jihad against Hausa in W. Africa (1804–8).	
1805 Nelson wins *Battle of Trafalgar*.	*Mehemet Ali* Pasha of Egypt.	
Austrians defeated at *Austerlitz* and make peace.		
1806 Death of *William Pitt*.		Beaufort scale of wind velocity.
Holy Roman Empire ends. Prussians defeated at *Jena*. Napoleon institutes *Continental System*. Confederation of Rhine formed.		Ingres: *Napoleon on his Throne*.

	Britain and Europe	US and Rest of the World	Culture/Technology
1807	British *Orders in Council*. Napoleon and Alexander I agree at *Tilsit* to divide Europe, east and west.	Colombia independence movement from Spain begins in *Venezuela* under *Miranda* and *Bolívar*.	Hegel: *Philosophy of History*. Fulton *paddle-steamer*.
1808	Spanish rising against French occupation; *Peninsular War* begins.	*James Madison* elected US President (1809–17). *Rum Rebellion* in NSW.	Gay-Lussac: law of gas expansion. 1808–32 Goethe: *Faust*.
1809	Renewed *war* with Austria, defeated at *Wagram*. British retreat from *Corunna*. France occupies *Papal States*.	*Macquarie* governor NSW. *Ecuador* independent.	Street lighting on Pall Mall, London.
1810	*Wellington* in command in Peninsular War.	Britain seizes *Cape Colony*. *Fulani* empire in Sokoto.	1810–14 Goya: *Disasters of War*.
1811	George III insane; prince regent installed (1811–20). British Orders in Council produce economic depression; *Luddite riots* against machines.	*Mehemet Ali* overthrows *Mamelukes* in Egypt. Rules with French advisers till 1849. *Ashanti* king *Osei Bonsu* crushes revolts and consolidates kingdom. Battle of *Tippecanoe*: US settlers defeat Indians. *Paraguay* independent.	Krupp factory at Essen. *Avogadro*: Law of atomicity of gases (Italy).
1812	*Napoleon* marches on *Russia*. Grand Army *retreats* from Moscow. British PM *Spencer Perceval* assassinated. *Liverpool* PM (1812–27).	*War of 1812*: US against Britain (1812–14).	Grimm: *Fairy Tales*. 1812–13 Byron: *Childe Harold*.
1813	*Battle of Leipzig*: allies defeat Napoleon. English East India Co. loses monopoly.		Davy: *Elements of Agricultural Chemistry*. Jane Austen: *Pride and Prejudice*.
1814	Napoleon abdicates to *Elba*. Louis XVIII (1814–24) and Ferdinand VII (1808, 1814–33) restored to France and Spain. *Jesuits* re-established. *Congress of Vienna* convened.	*Treaty of Ghent* ends war.	Machine *press* invented by Koenig prints *The Times*.
1815	Napoleon returns; *100 days*; raises army; defeated at *Waterloo*; to *St Helena*. 38 German States form *Confederation*. Alexander I forms *Holy Alliance*.	*Andrew Jackson* wins Battle of New Orleans. US *westward expansion* begins.	*Brighton Pavilion*. Davy: *miner's lamp*.
1816		*Congress of Tucumán* proclaims United Provinces of South America; against Spain.	Rossini: *Barber of Seville*. *Stethoscope* invented (France).
1817	British *Habeas corpus* suspended; income tax ended. 'Blanketeers' (cotton operatives) march in slump; leaders imprisoned.	*James Monroe* US President (1817–25).	Ricardo: *Principles of Political Economy*. John Keats: *Poems*. 1817–21 Weber: *Der Freischütz*.

	Britain and Europe	US and Rest of the World	Culture/Technology
1818		Shaka (d. 1828) forms Zulu kingdom.	Mary Shelley: Frankenstein.
1819	Second Factory Act in Britain. Peterloo 'massacre' (11 killed). Six Acts to suppress unrest.	Raffles founds Singapore. US purchases Florida. Bolívar forms Gran Colombia.	Schubert: Trout Quintet. Géricault: Le Radeau de la Méduse. Macadamized roads stimulate coach travel.
1820	Abortive risings in Portugal, Sicily, Germany, and Spain (Ferdinand suppresses Liberals). Cato Street Conspiracy against Government. George IV king (1820–30).	Missouri Compromise (no slavery in northern part of Louisiana Purchase); settlers beyond the Mississippi.	Percy B. Shelley: Prometheus Unbound. Scott: Ivanhoe. Constable: The Hay Wain. 1820–2 Saint-Simon: Système industriel.
1821	Famine in Ireland. Queen Caroline excluded from coronation Greek War of Independence starts.	Mexico independent.	Quincy: Confessions of an Opium-Eater.
1822	Castlereagh dies; Canning foreign secretary. Rapid industrialization of NW England and Lowland Scotland (textiles).	Liberia founded for US freed slaves.	Champollion deciphers Egyptian hieroglyphics (Rosetta Stone).
1823	Robert Peel initiates reform of prisons. Daniel O'Connell forms Catholic Association in Ireland.	Monroe Doctrine extends US protection to Spanish-American republics. First Anglo-Burmese War.	C. Macintosh rubberizes cotton (Scotland).
1824	British Combination Acts repealed to allow trade unions. Charles X in France (1824–30); repressive.	Peru independent.	Beethoven: Ninth Symphony. National Gallery founded.
1825	Nicholas I tsar of Russia (1825–55). Decembrist (rebel army officers) revolt suppressed.	Japan confirms laws of European exclusion. Java War (1825–30): Dutch establish firm control.	Stockton and Darlington railway.
1826	Mehemet Ali reconquers Peloponnese in Greek War.	Brazil independent. Britain establishes Straits Settlement: Penang, Malacca, and Singapore. Seku Ahmadu (Ahmad Lobbo) conquers Timbuktu.	J. F. Cooper: Last of the Mohicans.
1827	Britain and France send navies and destroy Turkish fleet at Navarino. Canning British PM; dies and succeeded by Wellington (1828–30).	France intervenes in Algeria.	Beethoven dies in Vienna.
1828	O'Connell (Catholic) elected to Parliament.		Brahmo Samaj (Hindu sect) formed in India.

	Britain and Europe	*US and Rest of the World*	*Culture/Technology*
1829	Full *Catholic emancipation* granted in Britain.	*Andrew Jackson* US President (1829–37).	*Braille* invented (France).
	Metropolitan Police formed.	*Western Australia* founded.	*Sewing-machine* (France).
	Treaty of Adrianople ends Russo-Turkish war; *Serbia* autonomous.		Stephenson: *Rocket*. Abolition of *suttee* in India.
1830	First *cholera epidemics* in Europe.	Mehemet Ali encourages revival of *Arabic culture*.	Stendhal: *Le Rouge et le noir*.
	Charles X deposed; *Louis Philippe* king (1830–48).	France takes *Algeria*.	Berlioz: *Symphonie fantastique*. Joseph Smith founds *Mormons*.
	William IV king of England (1830–7). *Lord Grey* PM.	*Colombia* and *Venezuela* independent.	Corot: *Chartres Cathedral*.
	Swing Riots in Britain; rural workers ruthlessly suppressed.		
	Polish rebellion suppressed.		
	Belgium independent.		
1831	Major *cholera* epidemic in Britain.		Delacroix: *La Liberté guidant le peuple*.
	Mazzini founds *Young Italy Movement*.		*Darwin* begins voyage on *Beagle*.
	Greece gains independence.		Bellini: *Norma*.
1832	First *Reform Act* in Britain.	*Black Hawk* war in US. His defeat leads to *Trail of Tears*; forcible settlement of Indians in *Oklahoma*.	Morse invents *code*.
	Industrialization in Belgium and NW France; railways.		
1833	*British Factory Act*: child labour regulated.	Abolition of *slavery* in British Empire.	*Oxford Movement* to restore Anglicanism.
1834	British *Poor Law* reformed; *workhouses* created.	Dorset *'Tolpuddle martyrs'* transported to Australia.	
	Creation of German *Zollverein*.		
	Civil War in Spain: Carlists against Liberals.		
	Peel's *Tamworth Manifesto*.		
1835	British *Municipal Reform Act*.	1835–7 *Great Trek* by Afrikaners in South Africa.	Donizetti: *Lucia di Lammermoor*.
	Railway boom (1835–7); Irish labour.		De Tocqueville: *Democracy in America*.
1836		*South Australia* becomes British Province.	*Neo-Gothic* architecture triumphant.
		Texas independent from Mexico.	1836–9 Chopin: *Twenty-four Preludes*.
1837	*Queen Victoria* (1837–1901).	*New Zealand* Association formed in London.	*Electric telegraph*.
		Rebellions in *Canada*.	
		Alafin Atiba rules *Oyo empire*, Nigeria (c.1837–59).	
1838	*People's Charter* drawn up in London for universal suffrage.	*Myall Creek* Massacre NSW.	*Daguerreotype* photograph.
		Battle of Blood River South Africa.	

	Britain and Europe	US and Rest of the World	Culture/Technology
1839	*Anti-Corn Law League* founded in Manchester.	First *Opium War* in China (1839–42).	*Fox Talbot* invents negative-positive photo.
	Chartist Movement in Britain (riots).	*Mehemet Ali* defeats Ottoman Turks.	Faraday: theory of *electromagnetism*.
		Tanzimat Reform in Ottoman Empire.	E. A. Poe: *Tales of the Grotesque and Arabesque*.
1840	Napoleon's ashes return to Paris.	Canadian Provinces *Act of Union*.	First *bicycle* in Scotland.
		Treaty of Waitangi, NZ: Maori and settlers.	*Penny post* in Britain.
		Mzilikazi leads *Ndebele* people into Transvaal and Mashonaland.	Proudhon: *On Property*.
			Schumann: *Lieder cycles*.
		Matabeleland founded by Ndebele.	
1841	*Robert Peel* British PM (1841–6).	*Said ibn Sayyid* makes *Zanzibar* his capital.	
		Britain takes *Hong Kong*.	
1842	*Income tax* reintroduced in Britain.	Reforms in *Japan*.	Verdi's *Nabucco* encourages Italian *nationalism*.
	British *Mines Act*: no women or children below 10 years.	*US/Canadian border* agreed.	Gogol: *Dead Souls*.
		Treaty ports in China (Treaty of Nanjing).	Balzac begins *Comédie humaine*.
		France occupies *Tahiti*.	Tennyson: *Morte d'Arthur*.
		1842–3 France occupies *Guinea* and *Gabon*.	
1843	Rural riots in Wales (*Rebecca*).	Britain conquers *Sind*.	Joule: *theory of thermodynamics*.
	Free Church of Scotland formed in protest against established Presbyterian Church.		I. K. Brunel: *SS Great Britain*: first iron/screw ship.
1844	*Bank Charter Act* regularizes British banking.		Dumas: *Count of Monte Cristo*.
	Rochdale Co-operative Society formed.		Turner: *Rain, Steam, Speed*.
	Royal Commission on Public Health.		Kierkegaard: *Concept of Dread*.
1845	*Potato famine* in Ireland; 1 m. die; 8 m. emigrate.	US annexes *Texas*.	Engels: *Condition of the Working Class in England*.
	Second Railway boom in Britain (1845–7).	*Sikh Wars* (1845–8/9).	Disraeli: *Sybil*.
			1845–9 Daumier: *Les Gens de justice*.
			Galvanized corrugated iron patented (England).
1846	*Corn Laws* repealed; Peel resigns; Lord *John Russell* PM (1846–52).	Second *Xhosa War*, Southern Africa.	
		Mexican–US War (1846–8).	
1847	*Il Risorgimento* in Italy.		Charlotte Brontë: *Jane Eyre*.
	Factory Act in Britain: 10-hour day.		Emily Brontë: *Wuthering Heights*.

	Britain and Europe	US and Rest of the World	Culture/Technology
1848	Charter *rejected*. British *Public Health Act.*	*Gold* discovered in *California.*	Liszt at Weimar.
	Revolutions in Europe and Ireland.	*Irish emigrants* to US.	*Communist Manifesto.*
	Second French Republic.	Abolition of *slavery* in French West Indies.	
	Cholera in Europe.	Britain annexes *Punjab* and subdues *Sikhs.*	
1849	Short-lived Roman Republic (Pope in exile).	California *gold-rush.*	*Safety-pin* patented.
	Revolutions *suppressed*; Garibaldi defeated in Italy and *Pope restored*; Austria defeats Hungarian nationalists.		
	Leopold II restored in Tuscany; repressive regime.		
1850		*Taiping Rebellion* in China; leader Hong Xiuquan (Hung Siu-tsuen); egalitarian beliefs; weakens Qing (Ch'ing) dynasty.	Millet: *Le Semeur.*
			Courbet: *L'Enterrement à Ornans* (salon scandal).
			Nathaniel Hawthorne: *Scarlet Letter.*
			Kelvin: *law of conservation of energy.*
1851	Fall of French Second Republic; coup by *Louis Napoleon*; plebiscite approves.	*Gold* in Australia; settlers moving into *Victoria.*	Paxton: Crystal Palace.
			Great Exhibition in Crystal Palace.
1852	*Aberdeen* British PM (1852–5).	Second *Anglo-Burmese War.*	W. H. Hunt: *Light of the World.*
	Napoleon III founds French *Second Empire.*	1852–6 *Livingstone* crosses Africa.	
	Cavour premier of *Piedmont.*	*Transvaal* independent.	
1853	*Gladstone's* first budget.	First *railways* and *telegraph* in India.	*Hypodermic syringe* (France).
	Fighting in *Crimea*; Russia against Turkey.	*Gadsden purchase* by Mexico (US).	Verdi: *La Traviata.*
		Russia completes conquest of Kazakhstan.	
1854	*Crimean War* develops; France and Britain join Turkey; *Siege of Sebastopol*; Nightingale to hospital at *Scutari.*	US forces *Japan* to end *isolation* (Treaty of Kanagawa).	Catholic dogma of *Immaculate Conception.*
	British *Civil Service* reformed.	*Eureka Rebellion* in Victoria.	
		Tukulor empire in Africa (al-Hajj Umar 1795–1864).	
		US *Republican Party* founded.	
		France annexes *Senegal.*	
1855	*Palmerston* British PM (1855–8).	*Railways* in South America: Chile (1851), Brazil (1854), Argentina (1857).	*Telegraph* news stories of Crimean War.
			Walt Whitman: *Leaves of Grass.*
			Courbet: *L'Atelier.*
			1855–6 Mendel discovers laws of *heredity.*

	Britain and Europe	US and Rest of the World	Culture/Technology
1856	*Peace of Paris* ends war; *Danube* open; Black Sea closed to warships.	Britain annexes *Oudh*. *1856–60* Second *Chinese Opium War*; Anglo-French force sacks *Summer Palace* and Peking; China open to *European trade*; Japanese trade agreement.	*Synthetic colours* invented.
1857	*Mazzini* hugely popular in London; *Italian nationalism* growing.	*1857–8 Indian Mutiny; atrocities* on both sides.	Baudelaire: *Fleurs du mal.* Flaubert: *Madame Bovary.* Trollope: *Barchester Towers.*
1858	Lionel Rothschild first *Jewish MP*. Government of *India* transferred from East India Co. to *Crown*.	*Burton* and *Speke* discover Lake *Tanganyika.* *Fenians* founded in US.	*Lourdes* miracles reported.
1859	Battles of *Solferino* and *Magenta*; French support Italians against Austria.	John Brown at *Harper's Ferry*.	J. S. Mill: *On Liberty.* Darwin: *Origin of Species.* *Oil* pumped in Pennsylvania.
1860	*Garibaldi's 1000*; Nice and Genoa to France.	*1860–70 Taranaki Wars* in New Zealand; *Maori* against *settlers*. *Lincoln* elected US President (1861–5); S. Carolina secedes from Union.	H. Bessemer mass production of steel. Huxley defends theory of evolution.
1861	*Victor Emmanuel* king of Italy (1861–78). Death of *Prince Albert*. Russia abolishes *serfdom*.	Bombardment of *Fort Sumter* leads to *Civil War* in US (1861–5). First Battle of *Bull Run*.	George Eliot: *Silas Marner.* *Siemens* developing *open-hearth steel* production.
1862	*Bismarck* minister-president of *Prussia*.	France annexes *Cochin-China*.	H. Spencer: *First Principles* (sociology). Victor Hugo: *Les Misérables.*
1863		*Battle of Gettysburg*; Union victory. *Emancipation* of US slaves. French protectorate of *Cambodia*.	Manet: *Déjeuner sur l'herbe* (salon scandal). Maxwell: *theory of electromagnetism.*
1864	Russia suppresses *Polish revolt*. First *Socialist International* in London (*Karl Marx* organizes). H. Dunant founds *Red Cross*.	French install *Maximilian* emperor of *Mexico* (shot 1867). *Sherman* march through *Georgia*.	Dickens: *Our Mutual Friend.* Jules Verne: *Voyage to Centre of the Earth.*
1865	*Earl Russell* British PM (1865–6).	*1865–70 Paraguayan War*: Brazil, Argentina, and Uruguay against Paraguay; ends in *disaster* for the latter. *Lee* surrenders at *Appomattox*. Lincoln assassinated; *Andrew Johnson* US President (1865–9).	Wagner: *Tristan und Isolde.* Lewis Carroll: *Alice's Adventures in Wonderland.* Pasteur publishes theory of germs causing disease.
1866	*Lord Derby* British PM (1866–8); *Second Reform Bill*. Prussia defeats Austria at *Sadova* (Seven Weeks War).	*1866–73 Livingstone* third journey in *Africa*. US *Reconstruction* Laws. US *Fourteenth Amendment* to Constitution.	Dostoevsky: *Crime and Punishment.* Swinburne: *Poems and Ballads.*

	Britain and Europe	US and Rest of the World	Culture/Technology
1867	Prussia forms *North German Confederation*; Austria forms *Austro-Hungarian* empire. *Fenian rising.* *Second British Reform Act.*	US purchases *Alaska* from Russia. *British North America Act* (Canada Dominion). *Meiji Restoration* in Japan; end of shogunates.	Karl Marx: *Das Kapital.* Ibsen: *Peer Gynt.* *Japanese art* arrives in West (Paris World Fair).
1868	*Gladstone* British PM (1868–74). British *TUC* formed.		W. Collins: *The Moonstone.* *Helium* discovered. First *refrigeration* ship. *Traffic signal* installed at Westminster, London.
1869	*Irish Church* disestablished.	Wyoming grants *women suffrage.*	*Suez Canal* opens. *Transcontinental Rail* (US). *Liquefaction of gases* (Andrews).
1870	Gladstone's *Irish Land Act.* British *elementary education.* Franco-Prussian War: Napoleon III defeated at *Sedan;* dethroned and exiled. *Papal Rome* annexed by Italy.	*Standard Oil* Company in US.	Doctrine of *papal infallibility.*
1871	French *Third Republic* suppresses *Paris Commune* and loses *Alsace-Lorraine* to *Second German empire.* British *trade unions* gain legality. *Bismarck* opposes Catholic church in '*Kulturkampf*' and establishes secular school system.	Stanley finds *Livingstone* at Ujiji. *Vancouver* and *British Columbia* join Canada. *Ku-Klux-Klan* suppressed in US.	
1872	*Scottish Education Act. Secret ballot* in Britain. British *Coal Mine Act.*		*Cézanne* settles in Provence. *Air brakes* patented (Westinghouse). Dewar: *vacuum flask.*
1873	*Marshall Mac-Mahon* President of France (1873–9).	Feudalism suppressed in Japan; modernization begins. 1873–1903 *Acheh War:* Dutch suppress revolt in *Sumatra.*	Monet: *Impression: Sunrise.* Cézanne: *La Maison du pendu à Auvers.* 1873–7 Tolstoy: *Anna Karenina.* Remington *typewriter.*
1874	*Disraeli* forms British government (1874–80). *French Factory* act bans children under 12.	British Colony of *Gold Coast.* US Reconstruction *collapsing:* southern States impose *racist* legislation.	First *Impressionist Exhibition:* Monet, Sisley, Renoir, Pissarro, Degas, Cézanne.
1875	Britain buys control of *Suez Canal. Public Health Act; Artisans Dwelling Act.*	*New Zealand* Parliament formed. *Savorgnan de Brazza* to Congo.	Bell patents *telephone.* Bizet: *Carmen.*
1876	Turkish *massacre* of Bulgarians; European outcry.	Victoria *Empress of India.* Battle of *Little Bighorn* (Custer's death). *Indian nationalism* growing.	*Plimsoll line* for ships. Brahms: *First Symphony.*

Britain and Europe	US and Rest of the World	Culture/Technology
1877	*Russo-Turkish War*; Russians threaten Constantinople; ended by *Treaty of San Stefano* (March 1878).	Last *Xhosa War* in South Africa. Britain annexes *Transvaal*. *Satsuma Rebellion* in Japan suppressed.	*Gramophone* (Edison). Tchaikovsky: *Swan Lake*. c.1877 *Microphones* developed by Berliner, Edison, and Hughes.
1878	*Congress of Berlin* settles Balkan crisis. *Salvation Army* created in Britain. *Serbian* independence.	Britain gains *Cyprus*. 1878–80 Britain fights *Second Afghan War*.	Gilbert and Sullivan: *HMS Pinafore*.
1879	*Dual Alliance*: Germany and Austria-Hungary. *Michael Davitt* forms *Irish Land League*; Parnell first President.	*Zulu War* in Africa. 1879–83 *War of Pacific*: Peru and Bolivia against Chile (backed by Britain) for saltpetre; Bolivia *land-locked*.	First *tramways* (Berlin).
1880	*Gladstone's Midlothian campaign* wins election.	1880–1 First *Boer War*. Britain releases *Transvaal* and *Orange Free State*.	*Swan* perfects *carbon-filament lamp*. Development of *seismograph*. Rodin: *La Porte de l'Enfer*.
1881	Irish *Land* and *Coercion Acts*. Assassination of *Alexander II*. *Jewish pogroms* in eastern Europe. *Alexander III* (1881–94).	1881–98 *Mahdi Holy War* in Sudan.	Nietzsche: *Aurora*. International *Electricity Exhibition* Paris.
1882	*Triple Alliance*: Germany, Austria, Italy. *Phoenix Park murders*, Dublin. Jules Ferry *'Loi scolaire'* secularizes French schools.	British occupy *Egypt*. French protectorate of *Tonkin*.	First *generating station* (New York). Manet: *Folies-Bergère*.
1883	*Plekhanov* founds Russian Marxist Party. Prussian *Social Security laws* (1883–7).	*Jewish* immigration to Palestine (Rothschild Colonies). Germany acquires *SW Africa*.	
1884	*Dreikaiserbund* renewed. *Third British Reform Act*. *Berlin Conference* on West African colonies.	1884–5 *Sino-French* war.	*Rayon* artificial fibres (France).
1885	*Lord Salisbury* British PM (1885–6). *Redistribution Act*.	Mahdi takes Khartoum; death of *Gordon* (Jan.). *Canadian Pacific* complete. All *Burma* occupied by British. Gold in Transvaal. Belgian Congo under *Leopold II*.	1885–98 Degas: *Femmes à leur toilette*. *Motor cycle* (Daimler). *Motor car* (Benz). *Pasteur* anti-*rabies* vaccine. Zola: *Germinal*. Rimbaud: *Les Illuminations*.
1886	*Gladstone* British PM but resigns over Irish Home Rule. *Joseph Chamberlain* forms *Liberal Unionists*. *Salisbury* PM (1886–92). *Anti-Semitism* increasing throughout Europe.	*Slavery* ends in *Cuba*. *Tunisia* under French protectorate. *Royal Niger Co.* Charter. First *Indian National Congress* meets.	

	Britain and Europe	US and Rest of the World	Culture/Technology
1887	Victoria *Golden Jubilee*. Industrial unrest in UK; *Trafalgar Square riots*. French General Boulanger fails to gain office.	British *East Africa Co.* Charter.	Conan Doyle: *Sherlock Holmes* stories. *Radio waves* (Hertz).
1888	British *match-girls* strike. *County Councils* formed. *Wilhelm II* emperor of Germany (1888–1918).	French *Indo-China* established. *Slavery* ends in Brazil.	*Pneumatic tyre* (Dunlop). *Van Gogh* to Arles.
1889	*British South Africa Co.* chartered. *London Dock strike.* Second *Socialist International*.	Japanese *Meiji Constitution*. Italy takes *Somalia* and *Ethiopia*. *Oklahoma* Indian territory opened for *settlers*.	*Eiffel Tower*, Paris. *Linotype* printing machine (Germany). 1889–93 Verdi: *Falstaff*.
1890	Support for Boulanger wanes. *Parnell* resigns (*divorce* scandal). *Bismarck* dismissed.	Australian *Maritime strike*. *Battle of Wounded Knee*.	Death of *Van Gogh*. Mascagni: *Cavalleria rusticana*.
1891	*Social Democrats* in Germany adopt *Kautsky's* Erfurt programme. *Trans-Siberian Railway* begun.	*Shearers'* Strike in Australia. Australian demands for trade protection and unification. *Young Turk* movement founded.	*Gauguin* to Tahiti. Toulouse-Lautrec: *Le Bal du Moulin-Rouge*.
1892	*Sergei Witte* Russian finance minister (1892–1903); modernizes finance. *Alexander III* forms alliance with France. Gladstone's *Fourth Ministry*. First *Labour MP*. French *Factory Act*: 11-hour day for women and children aged 13–16.		Maeterlinck: *Pelléas et Mélisande*.
1893	*Second Irish Home Rule* Bill rejected by Lords. *Independent Labour Party* formed.	*South Africa Co.* of Rhodes wages *Matabele War*. France annexes *Laos* and *Ivory Coast*. *Women's suffrage* in New Zealand.	Tchaikovsky: *Sixth Symphony*. Dvořák: *New World Symphony*.
1894	French President *Carnot* assassinated. *Nicholas II* tsar (1894–1917). *Lord Rosebery* PM (1894–5). *Death duties* in Britain.	*1894–5 Sino-Japanese War*; China overwhelmingly defeated. *Pullman strike* in US.	Kipling: *Jungle Book*. Escalator *lifts* (US).
1895	*Lord Salisbury* PM (1895–1902); strongly imperialist government.	*Cuban rebellion* begins. Japan takes *Taiwan* (Formosa). In Transvaal Kruger seeks to exclude *uitlanders*; *Jameson Raid* (1896).	Lumière Brothers: *cinema*. *X-rays* discovered (Röntgen). Wilde: *Importance of Being Earnest*. Gillette *safety-razor*.
1896	*Zionism* founded by Herzl. First *Olympic Games* in Athens. Kaiser Wilhelm supports *Kruger*.	Italian defeat at *Adowa*. French occupy *Madagascar*.	T. Hardy: *Jude the Obscure*. *Radioactivity* of uranium.

Britain and Europe	US and Rest of the World	Culture/Technology
1897 Victoria *Diamond Jubilee.* *Dreyfus Affair* in France.	British destroy *Benin City.*	J. J. Thomson: the *electron.* Bergson: *Matter and Memory.* *Compression ignition* engine (Diesel). *Monotype printing* (US). *Aspirin* marketed.
1898 German *naval expansion* rapid under *Tirpitz.*	*Spanish-American War:* Spain loses *Cuba, Puerto Rico,* and *Philippines.* British conquer Sudan (*Omdurman* and *Fashoda Incident*). *Hundred Days Reform* in China defeated by *Empress.* *Curzon* viceroy in India. Sudan under *Anglo-Egyptian protectorate.*	Zola: *J'accuse.* Curies: *radium.*
1899 Second *Boer War* (1899–1902); strong Liberal objections. Dreyfus pardoned.	British disasters in South Africa; relief of *Mafeking.* *Boxer Rising* in China.	
1900 British 'Khaki' Election. *Labour Representation Committee* formed. *Boer War concentration camps* arouse European criticism.	First *Pan-African* Conference.	First *Zeppelin.* Planck: *quantum theory.* Freud: *Interpretation of Dreams.* Conrad: *Lord Jim.* Puccini: *Tosca.* First agricultural *tractor.*
1901 *Death of Victoria,* end of an era. *Edward VII* (1901–10).	Commonwealth of Australia formed. British protectorates in *Nigeria.* US President *McKinley* assassinated. *Theodore Roosevelt* President (1901–9).	*Hearing-aid* patented (US). Strindberg: *Dance of Death.* 1901–4 Picasso: *Blue Period.*
1902 *Balfour* British PM (1902–5). *Education Act* creates *secondary schools.* German investment in Ottoman empire.	Peace of *Vereeniging* ends Boer War. *Anglo-Japanese* Alliance.	Gide: *L'Immoraliste.*
1903 *Bolshevik/Menshevik* split. *Pogroms* in Russia. *Suffragette movement* begins (WSPU).	Mass European *emigration* to US. *Panama Zone* to US. Britain takes *Sokoto;* ends Fulani empire.	*Electrocardiograph.* Wright Brothers' *flight.* Henry James: *The Ambassadors.*
1904 Franco-British *Entente Cordiale.*	1904–5 *Russo-Japanese War;* Russian *fleet* sunk.	Chekhov: *Cherry Orchard.* *Photoelectric* cell.

	Britain and Europe	US and Rest of the World	Culture/Technology
1905	Germany opposes French expansion in *Morocco*; *international crisis* to be resolved by conference at *Algeciras* (1906). First *Russian Revolution*. Liberal government in Britain; *Campbell-Bannerman* PM (1905–8).	Japanese protectorate of Korea.	R. Strauss: *Salome*. Einstein: *Special theory of relativity*. Debussy: *La Mer*. Early brain surgery (Cushing in US).
1906	*Liberals* win election. *Russian Duma*. UK *Trade Disputes Act*. *Clemenceau* French PM (1906–9). French strikes suppressed.	Revolution in *Iran*; parliament granted. *Muslim League* founded in India.	Matisse: *Bonheur de vivre*. Vitamins discovered (Hopkins). 1906–7 Gorky: *The Mother*.
1907	*Anglo-Russian Entente*. *Labour Party* formed.	*New Zealand* Dominion.	G. B. Shaw: *Major Barbara*. Electric *washing-machine*. Diaghilev Ballet.
1908	*Bulgaria* independent. *Bosnia-Herzegovina* occupied by Austrians. Asquith British PM (1908–16).		*Borstal system* in UK. 1908–9 Mahler: *Das Lied von der Erde*.
1909	Lloyd George *'People's Budget'* rejected by Lords. *Old Age Pensions* begin. *Dreadnought* programme.	*Congo Free State* under Belgian parliament. *Young Turk* revolution in Turkey. US supports revolution in *Nicaragua*. *Oil-drilling* in Iran by British.	Model T car (Ford).
1910	British *constitutional crisis* over Lords. *George V* king (1910–36).	Union of *South Africa*. Japan annexes *Korea*.	E. M. Forster: *Howards End*. 1910–13 Russell: *Principia mathematica* (with Whitehead). Kandinsky: *Concerning the Spiritual in Art*. Brancusi: *La Muse endormie*. *Post-Impressionist* Exhibition, London.
1911	British *constitutional crisis* resolved; *Parliament Act* restricts Lords. *National Insurance* provides sickness benefits. Industrial unrest in UK.	*Amundsen* reaches *South Pole*. Chinese revolution; *Sun Yixian* (Sun Yatsen) establishes *republic* (1912).	Rutherford: *nuclear model of atom*. 1911–13 Stravinsky: *Sacre du printemps*.
1912	*Balkan Wars* 1912–13: Greece, Serbia, Bulgaria, and Montenegro against Ottoman Turks.	*Italy* conquers *Libya*. ANC formed in South Africa. *Titanic* sinks. French protectorate in *Morocco*. W. *Wilson* elected US President (1913–21).	Jung: *Psychology of Unconscious*. Schoenberg: *Pierrot lunaire*. A. Fournier: *Le Grand Meaulnes*. First *parachute* descent from aircraft. *Stainless steel*.

Britain and Europe	*US and Rest of the World*	*Culture/Technology*
1913 *Ulster Volunteers* formed.		*Geiger-counter* invented.
Treaties of *London* and *Bucharest*: Turkey loses European lands except Constantinople.		D. H. Lawrence: *Sons and Lovers*.
		Apollinaire: *Les Peintres cubistes*.
Crisis in Ireland.		1913–27 Proust: *A la recherche du temps perdu*.
British *suffragettes*.		
1914 Assassination at *Sarajevo* 28 June. Outbreak of *First World War*. BEF at *Mons*; first battle of *Ypres* and *Marne*. *Trench warfare*.	*Panama Canal* opens. *Egypt* British *protectorate*.	1914–21 Berg: *Wozzeck*.
Battle of *Tannenburg*; Hindenburg takes 100,000 Russians prisoner.		
1915 *Dardanelles/Gallipoli* (March 1915–Jan. 1916).	Japan imposes *21 Demands* on China.	Einstein: *General theory of relativity*.
Asquith forms *coalition*.		
Zeppelin raids.		
1916 Battles of *Verdun*, *Somme*, and *Jutland*.	*Mesopotamia* campaign. British disaster of *Kut-el-Amara*.	*Dada movement*.
Irish *Easter Rising*, Dublin.	1916–18 *Arab revolt* against Ottoman Turks.	
Lloyd George PM (1916–22).		
1917 *Russian Revolutions* ('Feb.' and 'Oct.'). *Nicholas II* abdicates.	US enters *First World War*.	Pound: *Cantos*.
Passchendaele.	British take *Baghdad*.	Kafka: *Metamorphosis*.
British *convoy system*.	*Balfour Declaration* on Palestine.	Lenin: *The State and Revolution*.
		New Orleans *jazz*: first recording.
		De Chirico: *Le Grand Métaphysique*.
1918 Second battle of *Marne*; tanks help final allied advance.	British take *Palestine* and *Syria*; battle of *Megiddo*. Ottomans make peace.	G. M. Hopkins: *Poems* (posthumous).
Armistice (Nov.).		Spengler: *Decline of the West*.
Women's *suffrage* UK.		
1919 *Spartacist* revolt *Germany*.	Arab rebellion in *Egypt* against British.	Elgar: *Cello Concerto*.
Sinn Fein in Ireland.		Alcock and Brown fly Atlantic.
Poland, Hungary, Czechoslovakia, Estonia, Lithuania, Latvia become republics.	*Amritsar Massacre* in India. *Prohibition* in US.	Keynes: *Economic Consequences of the Peace*.
Third International, Moscow.		
Versailles Peace Settlement.		
1920 *IRA* formed.	*Gandhi* dominates Indian Congress.	
League of Nations.	W. *Wilson's* health fails; Senate *rejects* Versailles.	
Weimar Republic threatened by *Kapp Putsch*.	*Marcus Garvey* in New York.	

CHRONOLOGY OF WORLD EVENTS

	Britain and Europe	US and Rest of the World	Culture/Technology
1921	Irish Civil War (1921–4).	1921–2 Washington Conference on disarmament.	C. Chaplin: The Kid.
		1921–6 Spanish war against Berbers of Rif.	Pirandello: Six Characters in Search of an Author.
		Chinese Communist Party founded.	
		King Faisal in Iraq.	
1922	USSR formed.	King Fuad installed in Egypt.	J. Joyce: Ulysses.
	Italian fascists march on Rome. Mussolini forms government.		T. S. Eliot: Waste Land.
	Bonar Law British PM (1922–3).		Max Weber: Society and Economy.
1923	German hyper-inflation.	Ottoman empire ends; Palestine, Transjordan, and Iraq to Britain; Syria to France.	S. Spencer: The Resurrection.
	French occupy Ruhr.		Le Corbusier: Vers une architecture.
	Munich putsch fails.		First talkie.
	Baldwin British PM.		
1924	First Labour government; MacDonald PM.	Northern Rhodesia British protectorate.	
	Death of Lenin.	US economy booming.	
	Dawes Plan.	Achimota College in Gold Coast.	
	Baldwin PM (1924–9).		
1925	Stresemann German foreign minister. Locarno Pact.	Reza Khan shah in Iran.	Hitler: Mein Kampf.
		Chiang Kai-shek launches campaign to unify China.	Discovery of ionosphere.
			F. Scott Fitzgerald: Great Gatsby.
			Surrealist Exhibition in Paris.
			B. Keaton: films.
			Eisenstein: Battleship Potemkin.
1926	Germany joins League of Nations.	France establishes Republic of Lebanon.	First television.
	British General Strike fails after nine days.	Emperor Hirohito in Japan (1926–89).	
	Gramsci imprisoned in Italy.		
1927	Stalin comes to power; Trotsky expelled from Party.	Execution of Sacco and Vanzetti in US causes international outrage.	BBC founded.
	Red Army develops in USSR.		
	British Trade Disputes Act restricts trade unions.		
	Lindbergh flies Atlantic.		
1928	Soviet collectivization.	Kellogg–Briand Pact for peace.	Yeats: Tower.
	Nazis and Communists compete in Germany.		Ravel: Bolero.
	Women over 21 enfranchised in UK.		Penicillin discovered.
	Stresemann dies.		W. Disney: Mickey Mouse.
			1928–9 W. Walton: Viola Concerto.

CHRONOLOGY OF WORLD EVENTS

	Britain and Europe	US and Rest of the World	Culture/Technology
1929	Trotsky exiled. *Lateran Treaty* in Italy. *Yugoslavia kingdom* under kings of Serbia. Second *Labour government* under MacDonald. *Great Depression begins.*	*Wall Street Crash.* *Young Plan* for Germany.	
1930	London *Round-Table Conferences* on India 1930–2.	Gandhi leads *Salt March* in India. *Revolution* in Brazil; *Vargas* president.	E. Waugh: *Vile Bodies.* W. H. Auden: *Poems.* N. Coward: *Private Lives.* *Turbo-jet engine* patented by Whittle. Planet *Pluto* discovered. 1930–1 Walton: *Belshazzar's Feast.* 1930–1 Empire State Building in New York.
1931	*Global depression* worsens. *Alfonso XIII* flees; *Spanish Republic* formed. British *National government* under MacDonald (1931–5).	*Jiangxi Soviet* in China. *New Zealand* independent. *Japan* occupies *Manchuria.*	
1932	British *Union of Fascists* formed.	Gandhi campaigns for Indian *Untouchables.* *14 m. unemployed* in US. *Ottawa Conference.* Kingdoms of *Saudi Arabia* and *Iraq* independent.	*Cockcroft and Walton* split atom. James Chadwick: *neutron.* First *autobahn,* Cologne–Bonn, opened.
1933	*Nazi Party* wins German elections. Hitler appointed *chancellor;* racist laws (anti-Semitic) in Germany.	F. D. Roosevelt US President (1933–45); *New Deal.* *Batista* president of *Cuba.*	Malraux: *La Condition humaine.* Lorca: *Blood Wedding.*
1934	*Night of the Long Knives* in Germany; Hitler purges all rivals, including *Ernst Röhm; Third Reich* formed. Stalin *purges* begin.	1934–5 China *Long March.* *Purified National Party* in South Africa with ideology of *apartheid.*	H. Miller: *Tropic of Cancer.* A. Christie: *Murder on the Orient Express.*
1935	Nuremberg Laws: German *Jews* lose citizenship. *Stresa Agreement.* *Baldwin* British PM.	*Philippines* self-government. Italy invades *Ethiopia.* Prempeh II *Asantehene* in *Gold Coast.*	
1936	*Edward VIII* abdicates; *George VI* (1936–52). L. Blum 'Popular Front' government in France. *Anti-Comintern Pact* (Japan and Germany). *Spanish Civil War* (1936–9). *Rhineland* reoccupied.	Japan signs *anti-Comintern Pact.* Japan invades *China;* war 1937–45.	Mao Zedong (Tse-tung): *Strategic Problems of Revolutionary War.* Dali: *Soft Construction with Boiled Beans: Premonition of Civil War.* BBC: first public TV service. A. J. Ayer: *Language, Truth, and Logic.* Prokofiev: *Peter and the Wolf.*

Britain and Europe	US and Rest of the World	Culture/Technology
1937 Chamberlain British PM (1937–40).	Arab/Jewish conflict in Palestine.	Picasso: Guernica.
		Orwell: Road to Wigan Pier.
	Chinese Civil War truce; massacre of Nanjing.	Photocopier patented US.
1938 Austrian Anschluss with Germany. Munich Crisis. Czechoslovakia cedes Sudetenland.	Pan-Africanism gaining strength in East and West Africa.	Fluorescent lighting US.
		Nylon patented US.
		G. Greene: Brighton Rock.
IRA bombings in England.		Marx Brothers: films (1929–46).
		Nuclear fission discovered.
1939 Nazi/Soviet Pact. Poland invaded.		Pauling: Nature of Chemical Bond.
Franco caudillo of Spain.		
Britain and France declare war on Germany.		
1940 Occupation by Germans of France, Belgium, the Netherlands, Norway, Denmark. British retreat from Dunkirk. Vichy government in France.		
Churchill PM (1940–5). German bombing of Britain (day, 1940; and night).		
1941 Balkans occupied. German invasion of Soviet Union; scorched earth policy; siege of Leningrad.	Italians expelled from Somalia and Ethiopia and Eritrea.	O. Wells: Citizen Kane.
	Lend-Lease by US.	Messiaen to Paris Conservatoire.
	Atlantic Charter (Churchill and Roosevelt).	
	Pearl Harbor: loss of Philippines and Pacific islands. US enters war. Malaya, Singapore, and Burma to Japan.	
1942 Dieppe raid disaster.	Midway Island (June) and El Alamein (Nov.) key battles.	Nuclear reactor built (Fermi).
Beveridge Report.	Brazil enters war.	Camus: L'Étranger.
Battle of Stalingrad; German army surrenders (1943).		
1943 Allied bombing of Germany. Battle of Kursk (4000 tanks).		Cousteau and Gagnan: aqualung.
		Sartre: Being and Nothingness.
Allies invade Italy; Mussolini deposed and Fascist Party ends.		Kidney machines.
		T. S. Eliot: Four Quartets.
1944 Normandy invasion; Paris liberated; NW Europe campaign. Arnhem disaster.	MacArthur begins reconquest.	Holmes: Principles of Physical Geography.
	1944–5 Burma recaptured.	
British Education Act.	Battle of Leyte Gulf (Oct.).	1944–6 Eisenstein: Ivan the Terrible.
Civil War in Greece (1944–9).	Burma Road to China reopened.	

	Britain and Europe	US and Rest of the World	Culture/Technology
1945	Yalta Agreement. War ends in Europe (May). United Nations formed. Labour wins British election; Attlee PM (1945–51). Potsdam Conference.	Death of Roosevelt; Truman US President (1945–53). Pan-African Conference (Manchester). Atomic bombs on Hiroshima and Nagasaki. War ends (Sept.). Wars of independence in Indo-China and Indonesia. Civil War in China (1945–9).	Olivier: Henry V. B. Britten: Peter Grimes. J. Pollock: abstract expressionist painting, New York.
1946	Cold War begins. In Britain National Health Service and National Insurance. Italian Republic formed.	Perón President in Argentina. Jordan independent.	
1947	Fuel, power, transport nationalized. Marshall Plan for Europe, rejected by Soviets. Puppet Communist States in eastern Europe.	India independent.	Transistor US. Genet: The Maids. Tennessee Williams: Streetcar named Desire.
1948	Berlin airlift.	Ceylon in Commonwealth. 1948–60 Malayan emergency. Apartheid legislation in South Africa. Gandhi assassination.	Brecht: Caucasian Chalk Circle.
1949	Comecon and NATO formed. Republic of Ireland formed. Communist regime in Hungary.	Truman's Point Four aid program. People's Republic in China.	Orwell: 1984. Manchester Mk I computer. Simone de Beauvoir: The Second Sex.
1950	Schuman Plan: France and Germany to pool coal and steel industries. Labour Party wins election in Britain; retains power.	1950–3 Korean War. China conquers Tibet.	Stereophonic sound (2-track tape; phonograph record, 1958). Ionesco: The Bald Prima Donna. Rashomon; West discovers Japanese film. First successful kidney transplant. F. Hoyle: Nature of Universe.
1951	Conservatives win election; Churchill PM (1951–5).	Colombo Plan. Anzus pact in Pacific. Cyprus bid for independence.	Stravinsky: Rake's Progress. Matisse: Vence windows.
1952	Britain tests atomic bomb. European Coal and Steel Community formed; Britain refuses to join. Elizabeth II queen (1952–).		Hemingway: Old Man and the Sea. Beckett: Waiting for Godot.

	Britain and Europe	US and Rest of the World	Culture/Technology
1953	Death of *Stalin*. Conservative government in UK *denationalize* iron (1952), steel, and road transport.	*Egyptian Republic* formed. 1953–7 *Mau Mau* emergency Kenya. *McCarthy* era in US. *Eisenhower* US President (1953–61). *Central African Federation* formed (1953–63). *Korean War ends*.	Crick and Watson: double helix structure of *DNA*. Dylan Thomas: *Under Milk Wood*. *Colour TV* service (US).
1954	Britain rejects French *European Defence Plan*, which collapses.	*Dien Bien Phu* falls. *Geneva* Conference on Vietnam. *Algerian revolt* begins. British troops withdraw from Egypt. *Nasser* in power.	*Fortran* in US.
1955	West Germany joins *NATO*. *Warsaw Pact* formed. *Messina Conference*. A. *Eden* British PM (1955–7).	*Bandung Conference* of Third World nations.	*Hovercraft* patented. Marcuse: *Eros and Civilization*. Teilhard de Chardin dies (1881–1955).
1956	*Twentieth Congress* of Soviet Communist Party; *Khrushchev* denounces Stalin. Britain and France *collude* with Israel on *Suez*. *Polish and Hungarian* revolts.	*Hundred Flowers* in China. *Castro and Guevara* land in *Cuba*. *Suez War*. *Morocco* independent. French colonies autonomous. *Civil War* in *Vietnam* begins.	Osborne: *Look back in Anger*. 1956–9 *Guggenheim Museum* New York. First commercial nuclear power stations (Britain 1956, US 1957).
1957	*Macmillan* British PM (1957–63). *Treaty of Rome*; *EEC* formed. Soviet *Sputnik* flight.	*Ghana* independent.	Robbe-Grillet: *Jealousy*. J. Kerouac: *On the Road*. 1957–60 L. Durrell: *Alexandria Quartet*.
1958	*Fifth French Republic*. *De Gaulle* president. Pope *John XXIII* elected. *Life peerages* in Britain.	*Great Leap Forward* in China. British *West Indies Federation* (1958–62). *French Community* (1958–61).	*Silicon chip* invented by Texas Instruments. Lévi-Strauss: *Structural Anthropology*. Achebe: *Things fall Apart*. *Fast breeder reactor* at Dounreay.
1959	*Conservatives* win British election. *EFTA* formed. *North Sea* natural gas discovered.	*Cuban revolution*. *Cyprus* joins Commonwealth.	G. Grass: *The Drum*. Ionesco: *The Rhinoceros*.
1960	*U2 spy plane* wrecks US/USSR summit. Soviet *technicians* withdrawn from China.	End of *Malayan emergency*. *Sharpeville Massacre* South Africa. *Belgian Congo* independent. *Vietnam War* (1960–75). *OPEC* formed. *Nigeria* independent.	*Lasers* built (US). Oral *contraceptives* marketed. Battery *razor*. North Sea *gas* flow.

Britain and Europe	US and Rest of the World	Culture/Technology
1961 *Berlin Wall* erected.	*Bay of Pigs.* *South Africa Republic.*	Britten: *War Requiem.* Manned *space flight.*
1962 *Vatican II* (1962–5). *Cuban missile crisis* raises war threat. *Commonwealth Immigrants Act.*	Cuban missile crisis. Britain grants independence to *Jamaica, Trinidad and Tobago,* and *Uganda.*	*The Beatles.* Satellite television. A. Warhol: *Marilyn.*
1963 *French veto* Britain's EEC bid. *Test-ban Treaty* in Moscow. *Douglas-Home* PM (1963–4).	*Kenya* independent. *J. F. Kennedy* assassinated (President 1961–3). *OAU* formed. *Eritrea War* begins.	Hitchcock: *The Birds.* British *National Theatre* at Old Vic.
1964 Labour win British election; *H. Wilson* PM (1964–70). Khrushchev ousted by *Brezhnev* (1964–82). *First Race Relations Act in Britain.*	*Civil War* in the *Sudan.* US *Civil Rights Act* under President *Johnson* (1963–9). *PLO* formed.	*Word processor.*
1965 *De Gaulle* re-elected President. Wilson fails to solve *Rhodesia* crisis.	*UDI* in *Rhodesia.* *Indo-Pakistan War.* Military takeover in *Indonesia.*	Hoyle and Fowler: *Nucleosynthesis in Massive Stars and Supernovae.* H. Pinter: *The Homecoming.* J. Orton: *Loot.*
1966 *Labour* gains bigger majority.	*Cultural Revolution* in China (1966–8).	Chagall decorates Metropolitan Opera, New York.
1967 Economic problems result in *devaluation of pound.* EEC becomes EC. De Gaulle *vetoes* Britain's second bid to enter. *Abortion legalized* in Britain.	*Biafra War* (1967–70). *Indira Gandhi* Indian PM (1966–77, 1980–4). *Six-Day War* Israel.	Antonioni: *Blow-Up.* First *heart transplant* (Barnard).
1968 Warsaw Pact invade *Czechoslovakia.* *Salazar* resigns. *Student protests* throughout Europe.	*Tet offensive* in Vietnam. Martin Luther *King shot.*	Bell and Hewish discover first *pulsar.*
1969 Irish troubles begin. British army to *Ulster.* *De Gaulle* resigns. *Pompidou* president. *Brandt* German chancellor.	*Nixon* US President (1969–74). *Sino-Soviet* frontier war. *Gadaffi* in power in *Libya.* *Nimeiri* in power in the *Sudan.*	*Open university* in UK. First *man on moon.* Knox-Johnson: *Non-stop circumnavigation of earth.* *Concorde* flight. *Woodstock* pop festival.
1970 Conservatives win British election; *Heath* PM (1970–4). German policy of *Ostpolitik.* *N. Ireland* riots; *IRA Provisionals* recruit.	*Allende* president in *Chile.* *Biafra War* ends.	Saul Bellow: *Mr Sammler's Planet.*
1971 N. Ireland *internment* policy. British *Industrial Relations Act.*		Visconti: *Death in Venice.* *Decimal* currency UK. Fellini: *Roma.*

CHRONOLOGY OF WORLD EVENTS

	Britain and Europe	US and Rest of the World	Culture/Technology
1972	British *miners* strike. *Uganda* Asian refugees to UK. *Direct rule* of N. Ireland.	*Bangladesh* formed after Indo-Pakistan War. *US/USSR* detente. *Amin* seizes power in *Uganda*. *SALT I* signed. *US recognizes* Communist China. *Allende* killed in *Chile*.	1972–3 Berio: *Concerto for Two Pianos*. 1973–6 M. Tippett: *The Ice Break*.
1973	*Denmark*, *Ireland*, and *UK* enter European Community. *Helsinki Conference*. Widespread *industrial unrest* UK.	*US withdraws* from Vietnam War. *OPEC* raises oil prices. *Yom Kippur War*: Israel against Arab States.	US *Skylab* missions. Schumacher: *Small is Beautiful*. CT *body-scanners*. *Microwave* cooking becoming popular.
1974	Three-day week. *Wilson* PM (1974–6). *IRA* bombing of mainland Britain. *Portugal* restores *democracy*. N. Ireland Assembly fails.	*Watergate scandal*; *Nixon* resigns. Cyprus invaded by Turkey. *Haile Selassie* deposed in Ethiopia.	
1975	Franco dies; *Juan Carlos* restored to Spain. *British Referendum* confirms EC. *Sex Discrimination Act*. *IMF* assists Britain.	*Angola* and *Mozambique* independent. End of Vietnamese War. *Khmer Rouge* in Cambodia. *Civil War* in Lebanon.	*Apollo* and *Soyuz* dock in space.
1976	*Callaghan* British PM (1976–9).	Death of *Mao Zedong*; *Gang of Four*. *Soweto* massacre.	
1977	Democratic elections in *Spain*. *Charter 77* launched. *Terrorist activities* in Germany and Italy.	*Deng Xiaoping* gains power in China.	*Pompidou Centre*, Paris.
1978	Pope *John Paul II* elected.	*Camp David Accord* (President Carter, Israel, and Egypt). *Boat people* leave Vietnam. Civil wars in *Chad* and Nicaragua.	I. Murdoch: *The Sea, the Sea*. First *test-tube baby* born.
1979	*European Parliament* direct elections. 'Winter of discontent' *strikes* in UK. *Devolution referenda* in Wales and Scotland fail. Conservatives win election; *Thatcher* British PM (1979–90).	Civil war in *El Salvador*. Shah of Iran deposed by *Khomeini*. *Iran hostage* crisis. *USSR* invades *Afghanistan*. *Pol Pot* deposed in Cambodia.	*Islamic fundamentalism* spreading. Coppola: *Apocalypse Now*.
1980	*Solidarity* in Poland; *Catholic Church* actively anti-Communist.	*Zimbabwe* independent. *Iran–Iraq War* (1980–8). US funds *Contras* in Nicaragua.	W. Golding: *Rites of Passage*. Anglican Church *alternative service book*.

	Britain and Europe	US and Rest of the World	Culture/Technology
1981	British *Labour Party* split; *SDP* formed. *Riots* in London and Liverpool. *British Nationality Act. Privatization* of public corporations. High unemployment. *Mitterrand* French President.	R. *Reagan* US President (1981–9). *Sadat* shot in Egypt.	Space shuttle *Columbia.* Pierre Boulez: *Répons. Microprocessor* in a variety of domestic appliances and gadgets.
1982	British *victory* over Argentina in *Falklands War.*	*PLO* takes refuge in *Tunis.* Israel invades *Lebanon. Famine* in *Ethiopia.*	*Cordless* telephone. *Laser* discs (*CDs*).
1983	*Thatcher* re-elected PM. *Cruise missiles* installed in UK and Germany; *peace movements* active.	US troops invade *Grenada. Civilian government* restored in Argentina.	R. Attenborough: *Gandhi.*
1984	*Miners'* strike UK (1984–5). *IRA* bomb attack on Cabinet at Brighton.	*Hong Kong* Agreement UK and China. *Assassination of Indira Gandhi.*	*Thames Barrier* completed. *Office practices* revolutionized by *Fax, word processors,* and *PCs. Genetic fingerprinting* in forensic science.
1985	*Gorbachev* general secretary of Soviet Communist Party; begins policy of *liberalization. Anglo-Irish Agreement.*	*Eritrean Liberation Front* continues war against Ethiopia. *Anti-apartheid* movement developing in South Africa. Military coup in the *Sudan.* Threat to *tankers* through Iran–Iraq War. Syria seeks to pacify Lebanon.	*Electronic culture* replacing literary.
1986	*Spain* and *Portugal* join EC. *Jacques Chirac* French PM; 'co-habitation' with President. British unemployment peaks at 3.5 m. *Single European Act.*	US bombs *Libya.*	*Chernobyl* disaster. *Wole Soyinka* gains Nobel Prize. *Musée d'Orsay* in Paris. Tarkovsky: *The Sacrifice.*
1987	*Stock-market crisis* (Oct.). Third *Thatcher* government.	*China* suppresses *Tibet* protests. US/USSR *INF Treaty* for intermediate missiles. Palestinian *intifada.*	
1988	SDP and Liberals merge to form *Liberal Democrats. Mitterrand re-elected.*	Iran–Iraq war ends. *Polisario* struggle against Morocco. PLO recognizes Israel.	Salman Rushdie: *Satanic Verses.* Hawking: *Brief History of Time.*
1989	*Berlin Wall* broken. Communist leaders deposed: Hungary, Poland, East Germany, Czechoslovakia, Bulgaria, Romania.	*Khomeini* dies. G. *Bush* US President (1989–93). *Tiananmen Square* massacre (June). *Namibia* independent.	France celebrates the 1789 Revolution. *Communist* ideology *collapsing* throughout eastern Europe.

	Britain and Europe	*US and Rest of the World*	*Culture/Technology*
1990	*East and West Germany* reunited. *Thatcher* deposed; *Major* becomes PM.	*Iraq* invades and annexes Kuwait. ANC talks with De Klerk. *Cold War* formally ended.	800 m. watch *World Cup*.
1991	Conflict begins in *Yugoslavia* Failed coup in *Soviet Union* followed by its break-up	US-led coalition in *Gulf War* (Jan.–Feb.). Iraq surrenders after setting fire to Kuwaiti oil-wells.	
1992	*Major* re-elected PM Fighting in *Bosnia*	*Famine* in eastern and southern Africa *Bill Clinton* elected US president (1993–).	*Olympics* in Barcelona
1993	*Czechoslovakia* divides into Slovakia and Czech Republic		

CHRONOLOGY OF SCIENTIFIC DEVELOPMENTS

MEDICAL SCIENCE

*c.*460 BC Hippocrates brings medicine from the realm of magic and the super-natural into that of natural phenomena (Greece)

*c.*160 AD Galen shows that arteries contain blood, not air, and founds *experimental physiology* (Greece & Italy)

*c.*1530 Paracelsus introduces *chemical treatment of disease* (Austria)

1625 *circulation of the blood* described by W. Harvey (Britain)

1676 presence of *microbes* first detected by A. van Leeuwenhoek (Holland)

1796 first effective *vaccine* (against smallpox) developed by E. Jenner (Britain)

1816 *monaural stethoscope* designed by R. Laënnec (France)

1825 *blood transfusion* successfully performed by J. Blundell (UK)

1842 *ether* first used as an anaesthetic by C. Long (US)

1860 *pasteurization* technique developed by L. Pasteur (France)

1863–4 *clinical thermometer* introduced by W. Aitken (UK)

1865 *germ theory of disease* published by L. Pasteur (France)

1867 first *antiseptic operation* performed by J. Lister (UK)

1885 *cholera bacillus* identified by R. Koch (Germany)

1891 concept of *chemotherapy* developed by P. Ehrlich (Germany)

1895 *X-rays* discovered by W. Röntgen (Germany)

1897 first synthetic *aspirin* produced by F. Hoffmann (Germany)

1898 medical properties of radiation discovered by P. and M. Curie (France)

1901 existence of *blood groups* discovered by K. Landsteiner (Austria)

1921 *insulin* isolated by F. Banting and C. Best (Canada)

1922 *tuberculosis vaccine* developed by L. Calmette and C. Guérin (France)

1928 *penicillin* discovered by A. Fleming (UK)

1929 *iron lung* designed by P. Drinker and C. McKhann (US)

1932 first *sulphonamide* developed by G. Domagk (Germany)

1938–40 *penicillin* isolated by H. Florey (Australia) and E. Chain (Germany)

1943 first *kidney machine* designed by W. Kolff (Holland)

1944 *DNA* determined as the genetic material of life by O. Avery (US)

1950 first successful *kidney transplant* performed by R. Lawler (US)

1953 *double helix structure of DNA* discovered by F. Crick (UK) and J. Watson (US)

1955 *ultrasound* successfully used in body scanning by I. Donald (UK)

1958 first *internal cardiac pacemaker* implanted by A. Senning (Sweden)

1960   *contraceptive pill* first available (US)

1967   first *heart transplant* performed by C. Barnard (S. Africa)

1971.   *CAT* (computerized axial tomographic) scanner developed by G. Hounsfield (UK)

1978   first *test-tube baby* born (UK)

1980   World Health Organization declares the *world free of smallpox* from 1 January 1980

1983   *HIV* or human immunodeficiency virus identified as responsible for causing Aids (US & France)

1984   *genetic fingerprinting* developed by A. Jeffreys (UK)

TELECOMMUNICATIONS REVOLUTION

1794   *semaphore telegraph* designed by C. Chappe (France)

1837   *electric telegraph* patented by W. Cooke and C. Wheatstone (UK)

1838   *Morse code* introduced by S. Morse (US)

1839   first *commercial telegraph line* installed in London (UK)

1848   Associated Press *news wire service* begins in New York (US)

1851   first *underwater cable* laid under the English Channel

1851   Reuters *news wire service* begins in London (UK)

1858   first *transatlantic cable* laid between Ireland and Newfoundland

1858   *automatic telegraph system* patented by C. Wheatstone (UK)

1872   *duplex telegraphy* patented by J. Stearns (US)

1875   *telephone* patented by A. Bell (UK/US)

1877   *phonograph* (sound-recording machine) developed by T. Edison (US)

1878   first *telephone exchange* opened in Connecticut (US)

1887   existence of *radio waves* demonstrated by H. Hertz (Germany)

1889   first *automatic telephone exchange* introduced by A. Strowger (US)

1891   *discs for recording sound* pioneered by E. Berliner (Germany)

1895   G. Marconi (Italy) demonstrates *radio transmission* using Hertz's equipment

1897   *cathode-ray tube* invented by F. Braun (Germany)

1898   machine for *magnetic recording of sound* patented by V. Poulsen (Denmark)

1901   G. Marconi (Italy) sends first *radio signals* across the Atlantic

1906   *triode valve* invented by L. de Forest (US)

1920   regular *public radio broadcasting* established in Britain and US

1926   *television* first demonstrated by J. Baird (UK)

1927   first *transatlantic telephone links* opened between London and New York

1928   *magnetic tape* introduced by F. Pfleumer (Germany)

1936   *black and white television* service introduced by the BBC (UK)

1953   *colour television* successfully transmitted in US

1954   first *transistor radio* developed by Regency Company (US)

1956	first *transatlantic telephone cable* was laid underwater between Scotland and Newfoundland
1956	*video tape recorder* introduced by A. Poniatoff (US)
1960	Echo and Courier satellites (US) launched to relay first *satellite telephone calls* between US and Europe
1966	*fibre-optic cables* for telephone links pioneered by K. Kao and G. Hockham (UK)
1968	*pulse code modulation system* installed in London (UK)
1970	*videodisc* introduced by Decca (UK) and AEG (Germany)
1975	first *domestic videotape system* (Betamax) introduced by Sony (Japan)
1980s	*teletext* system developed and introduced into Europe
1982	*compact disc* produced by Philips (Holland) and Sony (Japan)
1989	computers that run on *light pulses* rather than electricity developed by British Telecom (UK)

COMPUTER TECHNOLOGY

1642	machine for *adding and subtracting* designed by B. Pascal (France)
1674	machine for *multiplying and dividing* designed by G. Leibniz (Germany)
1834	*analytical engine* designed by C. Babbage (UK)
1849	*representation of logical events by symbols* enumerated by G. Boole (UK)
1855	*calculating engine* constructed by G. Schentz (Sweden) and exhibited at the Paris Exhibition
1889	*punched-card machine* patented by H. Hollerith (US) and used to tabulate the results of the US census
1924	Tabulating Machine Company of the US becomes International Business Machines (IBM)
1928	*magnetic tape* introduced by F. Pfleumer (Germany)
1943	code-breaking machine *Colossus* conceived by A. Turing (UK)
1945	*Electronic Numerical Integrator and Calculator* (ENIAC) designed by P. Eckert and J. Maunchly (US)
1947	*transistor* invented by J. Bardeen, W. Brattain, and W. Shockley at Bell Laboratories (US)
1948	first *computer, Manchester Mark I*, installed at Manchester University (UK)
1949–50	*printed electronic circuits* developed
1954	first *high-level programming language* (FORTRAN) published by J. Backus of IBM (US)
1958	first *integrated circuit* (or *silicon chip*) produced (US)
1962	Sinclair Radionics founded by C. Sinclair (UK)
1964	first *word processor* introduced by IBM (US)
1965	first *minicomputer* produced (US)
1968	Amstrad (UK) founded by A. M. Sugar to produce *microcomputers*
1969	*microprocessor* invented by E. Hoff of Intel (US)

1972	first *pocket calculator* introduced (UK)
1975	first *portable computer* produced by Altair (US)
1978	*magnetic discs* first introduced by Oyz (US)
1979	first *videotex* information system, Prestel, launched by British Telecom (UK)
1980s	term *fifth generation* coined to describe computers designed for parallel processing (Japan)
1981	*desktop microcomputer* (IBM-PC) introduced by IBM (US)
1983	ESPRIT (European Strategic Programme for Research and Information Technology) established by the EEC to coordinate the European Information Technology industry
1984	*mouse* (movable desktop pointing device) introduced by Apple Macintosh (UK)
1984	*Data Protection Act* passed in UK, requiring disclosure of personal computer-held information to anyone requesting it
1985	*transputer* manufactured by Inmos (UK)
1985	*optical fibres* first used to link mainframe computers (US)
1986	class of *superconductors* discovered by G. Bednorz and A. Mueller at IBM laboratories (Zurich), with potential for superconducting computers in the future
1988	improved optical discs for general computing manufactured by ICL (UK) and Tender Electronic Systems (US)
1989	RISC (Reduced Instruction Set Computer) chip developed by INTEL (US), performing at ten times the speed of previous chips

SPACE EXPLORATION

1903	theory of *rocket propulsion* published by K. Tsiolkovsky (Russia)
1923	theory of *interplanetary flight* published by H. Oberth (Germany)
1926	first *liquid-fuel rocket* launched by R. Goddard (US)
1942	first *long-distance rocket* designed by W. von Braun (Germany)
1944	V-2 rocket, forerunner of *space rockets*, first used (Germany)
1949	first *multi-stage rocket* launched in US
1957	a *satellite*, *Sputnik 1* (USSR), first put into earth orbit
1957	satellite *Sputnik 2* (USSR) carries the *first space traveller*, the dog Laika
1958	US follows USSR and first puts a satellite, *Explorer 1* (US), into orbit
1959	space probe *Luna 2* (USSR) reaches the Moon
1961	Y. Gagarin becomes the *first man in space* in *Vostok 1* (USSR)
1963	V. Tereshkova becomes the *first woman in space* in *Vostok 6* (USSR)
1964	space probe *Ranger 7* (US) takes close pictures of the Moon
1965	France becomes third nation to launch a satellite
1965	A. Leonov makes the first *walk in space* from *Voskhod 2* (USSR)
1966	first *moon landing*, by space probe *Luna 9* (USSR)

1966   N. Armstrong and D. Scott in *Gemini 8* (US) make the *first docking in space* with *Gemini Agena* target vehicle

1968   F. Borman, J. Lovell, and W. Anders orbit the Moon in *Apollo 8* (US)

1969   N. Armstrong and E. Aldrin make the *first manned lunar landing* in the lunar module *Eagle*, with M. Collins in the command module *Apollo 11* (US). Armstrong and Aldrin walk on the Moon.

1970   Japan and China become the fourth and fifth satellite-launching nations

1971   Britain becomes the sixth satellite-launching nation

1971   *Soyuz 10* and *Salyut 1* (USSR) dock in orbit to form the first space station

1973   *Skylab* (US) launched for a series of zero-gravity experiments

1975   *Soyuz 19* (USSR) and *Apollo 18* (US) dock in orbit for the Apollo–Soyuz Test Project

1976   *Viking 1* (US) lands on Mars and sends back information on conditions and signs of life. No conclusive evidence of life on the planet

1977   Space Shuttle *Enterprise* (US) makes the *first shuttle test flight*

1977   *Voyager 2* (US) leaves on a mission to fly by Jupiter (1979), Saturn (1981), Uranus (1986), and Neptune (1989)

1981   Space Shuttle *Columbia* (US) makes the first shuttle mission

1983   European Spacelab launched into orbit by the Space Shuttle *Columbia* (US)

1986   European space probe *Giotto* investigates Halley's Comet

1986   Space Shuttle *Challenger* (US) explodes killing seven astronauts and halting the shuttle programme

1990   European Space Telescope Hubble launched in order to obtain clearer stellar images

1992   COBE satellite provides evidence of *ripples* in the background radiation of the Universe

OXFORD

MORE OXFORD PAPERBACKS

This book is just one of nearly 1000 Oxford Paperbacks currently in print. If you would like details of other Oxford Paperbacks, including titles in the World's Classics, Oxford Reference, Oxford Books, OPUS, Past Masters, Oxford Authors, and Oxford Shakespeare series, please write to:

UK and Europe: Oxford Paperbacks Publicity Manager, Arts and Reference Publicity Department, Oxford University Press, Walton Street, Oxford OX2 6DP.

Customers in UK and Europe will find Oxford Paperbacks available in all good bookshops. But in case of difficulty please send orders to the Cash-with-Order Department, Oxford University Press Distribution Services, Saxon Way West, Corby, Northants NN18 9ES. Tel: 0536 741519; Fax: 0536 746337. Please send a cheque for the total cost of the books, plus £1.75 postage and packing for orders under £20; £2.75 for orders over £20. Customers outside the UK should add 10% of the cost of the books for postage and packing.

USA: Oxford Paperbacks Marketing Manager, Oxford University Press, Inc., 200 Madison Avenue, New York, N.Y. 10016.

Canada: Trade Department, Oxford University Press, 70 Wynford Drive, Don Mills, Ontario M3C 1J9.

Australia: Trade Marketing Manager, Oxford University Press, G.P.O. Box 2784Y, Melbourne 3001, Victoria.

South Africa: Oxford University Press, P.O. Box 1141, Cape Town 8000.

HISTORY IN OXFORD PAPERBACKS

Oxford Paperbacks' superb history list offers books on a wide range of topics from ancient to modern times, whether general period studies or assessments of particular events, movements, or personalities.

THE STRUGGLE FOR
THE MASTERY OF EUROPE 1848–1918

A. J. P. Taylor

The fall of Metternich in the revolutions of 1848 heralded an era of unprecedented nationalism in Europe, culminating in the collapse of the Hapsburg, Romanov, and Hohenzollern dynasties at the end of the First World War. In the intervening seventy years the boundaries of Europe changed dramatically from those established at Vienna in 1815. Cavour championed the cause of *Risorgimento* in Italy; Bismarck's three wars brought about the unification of Germany; Serbia and Bulgaria gained their independence courtesy of the decline of Turkey—'the sick man of Europe'; while the great powers scrambled for places in the sun in Africa. However, with America's entry into the war and President Wilson's adherence to idealistic internationalist principles, Europe ceased to be the centre of the world, although its problems, still primarily revolving around nationalist aspirations, were to smash the Treaty of Versailles and plunge the world into war once more.

A. J. P. Taylor has drawn the material for his account of this turbulent period from the many volumes of diplomatic documents which have been published in the five major European languages. By using vivid language and forceful characterization, he has produced a book that is as much a work of literature as a contribution to scientific history.

'One of the glories of twentieth-century writing.' *Observer*

Also in Oxford Paperbacks:

Portrait of an Age: Victorian England   G. M. Young
Germany 1866–1945   Gorden A. Craig
The Russian Revolution 1917–1932   Sheila Fitzpatrick
France 1848–1945   Theodore Zeldin

OPUS

*General Editors: Walter Bodmer, Christopher Butler,
Robert Evans, John Skorupski*

OPUS is a series of accessible introductions to a wide
range of studies in the sciences and humanities.

METROPOLIS

Emrys Jones

Past civilizations have always expressed themselves in great
cities, immense in size, wealth, and in their contribution to
human progress. We are still enthralled by ancient cities like
Babylon, Rome, and Constantinople. Today, giant cities abound,
but some are pre-eminent. As always, they represent the greatest
achievements of different cultures. But increasingly, they have
also been drawn into a world economic system as communications have improved.

Metropolis explores the idea of a class of supercities in the
past and in the present, and in the western and developing
worlds. It analyses the characteristics they share as well as those
that make them unique; the effect of technology on their form
and function; and the problems that come with size—congestion,
poverty and inequality, squalor—that are sobering contrasts to
the inherent glamour and attraction of great cities throughout
time.

Also available in OPUS:

The Medieval Expansion of Europe   J. R. S. Phillips
Metaphysics: The Logical Approach   José A. Benardete
The Voice of the Past 2/e   Paul Thompson
Thinking About Peace and War   Martin Ceadel

PAST MASTERS

General Editor: Keith Thomas

The *Past Masters* series offers students and general readers alike concise introductions to the lives and works of the world's greatest literary figures, composers, philosophers, religious leaders, scientists, and social and political thinkers.

'Put end to end, this series will constitute a noble encyclopaedia of the history of ideas.' Mary Warnock

HOBBES

Richard Tuck

Thomas Hobbes (1588–1679) was the first great English political philosopher, and his book *Leviathan* was one of the first truly modern works of philosophy. He has long had the reputation of being a pessimistic atheist, who saw human nature as inevitably evil, and who proposed a totalitarian state to subdue human failings. In this new study, Richard Tuck shows that while Hobbes may indeed have been an atheist, he was far from pessimistic about human nature, nor did he advocate totalitarianism. By locating him against the context of his age, Dr Tuck reveals Hobbs to have been passionately concerned with the refutation of scepticism in both science and ethics, and to have developed a theory of knowledge which rivalled that of Descartes in its importance for the formation of modern philosophy.

Also available in Past Masters:

Spinoza   Roger Scruton
Bach   Denis Arnold
Machiavelli   Quentin Skinner
Darwin   Jonathan Howard

OXFORD LIVES

Biography at its best—this acclaimed series offers authoritative accounts of the lives of men and women from the arts, sciences, politics, and many other walks of life.

STANLEY

Volume I: The Making of an African Explorer
Volume II: Sorceror's Apprentice

Frank McLynn

Sir Henry Morton Stanley was one of the most fascinating late-Victorian adventurers. His historic meeting with Livingstone at Ujiji in 1871 was the journalistic scoop of the century. Yet behind the public man lay the complex and deeply disturbed personality who is the subject of Frank McLynn's masterly study.

In his later years, Stanley's achievements exacted a high human cost, both for the man himself and for those who came into contact with him. His foundation of the Congo Free State on behalf of Leopold II of Belgium, and the Emin Pasha Relief Expedition were both dubious enterprises which tarnished his reputation. They also revealed the complex—and often troubling—relationship that Stanley has with Africa.

'excellent . . . entertaining, well researched and scrupulously annotated' *Spectator*

'another biography of Stanley will not only be unnecessary, but almost impossible, for years to come' *Sunday Telegraph*

Also available:

A Prince of Our Disorder: The Life of T. E. Lawrence
John Mack
Carpet Sahib: A Life of Jim Corbett   Martin Booth
Bonnie Prince Charlie: Charles Edward Stuart   Frank McLynn

OXFORD LETTERS AND MEMOIRS

Letters, memoirs, and journals offer a special insight into the private lives of public figures and vividly recreate the times in which they lived. This popular series makes available the best and most entertaining of these documents, bringing the past to life in a fresh and personal way.

RICHARD HOGGART

A Local Habitation
Life and Times: 1918–1940

With characteristic candour and compassion, Richard Hoggart evokes the Leeds of his boyhood, where as an orphan, he grew up with his grandmother, two aunts, an uncle, and a cousin in a small terraced back-to-back.

'brilliant . . . a joy as well as an education' Roy Hattersley -

'a model of scrupulous autobiography' Edward Blishen, *Listener*

A Sort of Clowning
Life and Times: 1940–1950

Opening with his wartime exploits in North Africa and Italy, this sequel to *A Local Habitation* recalls his teaching career in North-East England, and charts his rise in the literary world following the publication of *The Uses of Literacy*.

'one of the classic autobiographies of our time' Anthony Howard, *Independent on Sunday*

'Hoggart [is] the ideal autobiographer' Beryl Bainbridge, *New Statesman and Society*

Also in Oxford Letters and Memoirs:

My Sister and Myself: The Diaries of J. R. Ackerley
The Letters of T. E. Lawrence
A London Family 1870–1900   Molly Hughes